Stereo Sound for Television

Francis Rumsey

Focal Press
London and Boston

Focal Press
is an imprint of Butterworth Scientific

PART OF REED INTERNATIONAL P.L.C.

First published 1989

British Library Cataloguing in Publication Data
Rumsey, Francis
Stereo sound for television.
1. Television. Sound recording
I. Title
621'.389'3
ISBN 0 240 51288 X

Library of Congress Cataloging in Publication Data
Rumsey, Francis.
Stereo sound for television / Francis Rumsey.
p. cm.
Bibliography: p.
Includes index.
ISBN 0-240-51288-X
1. Television—Equipment and supplies. 2. Stereophonic sound systems. I. Title.
TK6643.R86 1989
621.388—dc20
89-35568

Typeset by Latimer Trend & Company Ltd, Plymouth
Printed and bound by Courier International Ltd, Tiptree, Essex

Stereo Sound for Television

Francis Rumsey

Foreword

Francis Rumsey, it has to be said, is courageous. Within this volume, which bears a deceptively simple title, lies more room for contention and enthusiastic debate than most politicians would dare to contemplate. I suspect that when many other issues that surround us in 1989 are long laid to rest, television sound staff in many parts of the globe will still discuss Rumsey's solutions to 'the TV stereo problem' far into the night.

By that time, of course, 'The Problem' will have become quantifiable to some extent: we shall know how big the TV stereo marketplace became. We shall know if the average family wanted surround sound at home with its films. We shall know which techniques of pickup and treatment proved more acceptable. We shall know if commercial considerations drove some companies to put mono through a black box and call it stereo . . .!

All of us who live by sound know how much stereo can enhance the product. The marriage of stereo to the small screen has demanded a great deal of enthusiasm, education and hard work from several dedicated groups of people in the broadcasting industry. Francis understands and effectively conveys much of the current knowledge of the subject together with some interesting personal insights for those who care to develop them further. There is certainly much room for personal experimentation and development in this area and *Stereo Sound for Television* will provide valuable additional material for such developments to prosper.

Jeff Baker
Head of Television Sound, BBC

Preface

Stereo Sound for Television is written for everyone involved with television sound, whether engineer, operator or manager. Although stereo sound is nothing new technologically, having been used in commercial recording for some thirty years, television is only just beginning to embrace it on a day-to-day basis, finding that stereo operational techniques from the music business need considerable modification where a picture and dialogue are concerned.

Stereo sound is likely to involve the television company in changes of operational practice, equipment, and transmission systems. For this reason anyone who makes decisions about such matters must be well informed, and it is to this end that *Stereo Sound for Television* endeavours to provide comprehensive coverage of every angle from psycho-acoustics to digital transmission systems. If one does not understand how stereo is perceived by the listener, especially in conjunction with a picture, it is difficult to appreciate why certain operational techniques are better than others, and why, for example, one might need to improve the acoustics of a control room.

From the operator's point-of-view, the first chapters will be more directly relevant than the later chapters, as they deal with practice rather than engineering matters, but the keen operator will be interested to know what happens after his part of the job is finished, and how the stereo signal eventually finds its way to the home. Engineers and planners involved with the purchase of equipment, and also the distribution and transmission of the stereo signal, will perhaps find the later chapters more relevant, but should not ignore the more creative problems of operators.

The subject matter is intended to be relevant internationally. Aspects of theory do not change from country to country, although the implementation of that theory may vary in relation to internal codes-of-practice. Where national variations exist (such as in the area of transmission systems) examples from the most common systems are taken. Where possible, *Stereo Sound for Television* should be read in conjunction with the particular codes-of-practice relevant in the reader's company or country, as this will provide additional detail regarding the specific implementation required.

In engineering terms, the coverage is unreservedly slightly biased towards digital technology, without ignoring analogue solutions, as digital audio does away with most of the phase, level and frequency response errors which plague accurate stereo reproduction. Digital audio has already found its way into video recording equipment (the M-II format, D-1 and D-2 digital video formats), giving more audio tracks with higher sound quality; distribution systems (the AES/EBU interface and stereo sound-in-syncs); and transmission systems (NICAM 728 and Dolby ADM), incorporated in both terrestrial and satellite operations. Although digital video is unlikely to be transmitted to the home before the end of the century, it is quite feasible that there will soon be an entirely digital chain for sound, right from the original recording through to the decoder in the home.

In considering the production of television programmes in stereo, one must not forget the viewer at home, as domestic listening conditions will be vastly different to the sanctuary of the control room. Stereo sound must be made equally worthwhile for the viewer with a small screen and loudspeakers on either side as it is for the enthusiast with a 26″ (61 cm) monitor and a hi-fi system. It might even be said that since there are likely to be more of the former, the sound should be tailored for this type of viewing. Many might say that it is impossible to reproduce realistic stereo on loudspeakers only two feet apart, but that is how many people will receive it, and thus the balancer is placed in the difficult position of having to decide whether to exaggerate stereo separation, making the result somewhat gimmicky for the serious listener, or whether to balance in 'natural stereo'. Dilemmas such as this require careful study of microphone technique and post-production methods, together with an understanding of psycho-acoustics.

Similarly, one must not forget the mono listener, as stereo sound will find its way only gradually into people's homes. In one respect the sound balancer will be in a similar position to that which vision operators found themselves in when colour television first appeared, in that instead of having to check for mono*chrome* compatibility it will be necessary to check for mono*phonic* compatibility. After a number of years of colour operation it was deemed unnecessary to carry on checking monochrome compatibility, and in the same way it will become unnecessary to check monophonic sound compatibility once sufficient people are receiving in stereo.

Stereo sound offers great potential for the enhancement of television, especially in the realm of advertising where greater impact can be achieved through the use of stereo effects. When it is done well, stereo sound is better than mono sound, but when it is done badly (and there are plenty of examples of this) it would be better if the programme had been in mono. There is no doubt that the incorporation of stereo sound into programme production will take more thought than is required for mono sound, but day-to-day programmes need not always involve more time and money if operators understand the principles and techniques involved. Prestigious productions involving international sales will warrant the extra time spent in post-production to make a truly exceptional stereo production, and it is becoming common for foreign TV companies to insist that sound is in stereo before they will consider the purchase of a series or drama.

There has been no attempt in this book to dictate operational practice, as this is an area for considerable contention and personal opinion, although a number of suggestions are made and the reader is left to decide for himself what is the most appropriate approach in a particular situation. With a background in the theory and principles of microphone technique the reader will be well equipped to make operational decisions.

Finally, the author would like to thank the Independent Television Association (UK) for their support, and the opportunity to develop stereo sound training courses, without which the proposal for this book might not have come about.

Francis Rumsey
February 1989

Contents

1

Directional perception, acoustics and psycho-acoustics

Directional perception

The way in which the ear and brain detect directional information in sound has been the subject of considerable research, and many theories have been proposed to explain the perceived differences between sounds in front of and behind us, above and below us, and to either side of us. As far as stereo sound reproduction is concerned, the question of up/down and front/back perception is far less important than the question of left/right perception, as one is dealing with two-channels of information reproduced in the horizontal plane, usually on two loudspeakers. It seems unlikely that this situation will change domestically in the foreseeable future, simply for practical reasons of space in the home, although domestic Dolby surround-sound decoders are beginning to become available.

It is important to understand the principles of directional perception before the application of these principles can be appreciated in such areas as stereo microphone technique and loudspeaker reproduction. Having appreciated these matters it will be seen that most stereo microphone techniques do not in fact reproduce the original sound field correctly at the ears, but merely produce an *impression* of directionality, and a more-or-less realistic soundfield. This is perhaps the most that can be hoped for in the varied surroundings of the domestic living room and the placing of a television set.

Left-to-right perception

Level cues

Acoustic principles hold that an object only begins to act as a suitable barrier to sound when it is larger than the wavelength of the sound in question. Once the wavelength of the sound is considerably shorter than the object's dimensions some attenuation will result on the 'shadowed' side of the object (that is the side furthest from the sound source). The object in question, as far as hearing is concerned, is the head, and its dimensions are such that the head does not begin to act as a significant barrier to sound until the frequency of the sound is above a few kilohertz. At low frequencies, the level difference between the ears due to this effect is largely insignificant, although level differences between the ears may result for other reasons such as standing waves in a room. At very high frequencies it can be seen that significant level differences will arise between the ears for a sound which is located off to one side of the head.

Another point to bear in mind when considering the potential for level differences between the ears for off-centre sounds is that although the direct sound level from a source decays with the inverse square of the distance (that is 6 dB per doubling in distance from the source), the distance between the ears will usually be very small compared with the distance from the source to the ears, and thus the level drop between the ears will also be very small. Also, it is possible that the listener will be in a relatively diffuse sound field where the effect of distance from the source on the perceived sound level is smaller than it is in the direct field.

Clearly then, one way in which sounds may be perceived to be located off-centre is by the brain's measurement of the level difference between the ears, as the nearer ear will tend to perceive a higher sound level than the distant ear. This effect increases the more off-centre a source is, and reaches a maximum at 90° off-centre. It is only pronounced at high frequencies.

Time and phase cues

Time differences and phase differences between the ears both result from the same phenomenon, which is that sound travels at around 340 m s^{-1} and thus will arrive at a different time for each ear if the source is off centre, only arriving at both ears simultaneously if the source is equi-distant from the two ears: that is in the centre plane. The result is usually known as a *phase* difference when continuous repetitive waveforms are discussed, as the time difference appears as a number of degrees phase shift in the waveform perceived at one ear when compared to that perceived at the other. It is usually more appropriate to refer to a *time-of-arrival* difference between the ears, as most real sounds are not simple repetitive waveforms, having transients, changes in timbre, and pauses.

It has been suggested that phase difference between the ears is the most important factor in determining direction at low frequencies, where level differences do not arise, and indeed if one listens to a sine wave reproduced from a loudspeaker then there will indeed be a phase difference in the signals received by the two ears if the loudspeaker is off centre. Whether this cue is important or not depends on its frequency, as once the frequency of the sound is above that at which half its wavelength equals the path-length difference between the ears, it is possible for the same phase difference to arise at a number of possible frequencies, thus confusing the brain. Also, at frequencies where the path-length difference between the ears amounts to exactly half a wavelength and its multiples there will be no phase difference at all. Luckily this confusion only begins to arise above roughly the same frequency as that at which the level difference cues begin to take over. In the cross-over region it is very difficult to locate continuous sound sources: witness the difficulty in locating the source of a 1 kHz sine wave.

As already stated, we rarely listen to continuous sine waves, and thus the measurement by the brain of phase difference becomes difficult with complex sound sources. More importantly, the measurement of the time difference between arrivals of the sound at each ear is used to determine the direction, and such a measurement can take place at any identifiable change in the sound, such as a transient, a pause, or a change in quality. The veracity of this is proven by the improved ability of the brain to detect directionality in sounds with a high transient content or a lot of changes in timbre, and the effect is largely independent of the frequency of the sound.

The time difference varies with the angle subtended by the source to the forward axis: see Figure 1.1.

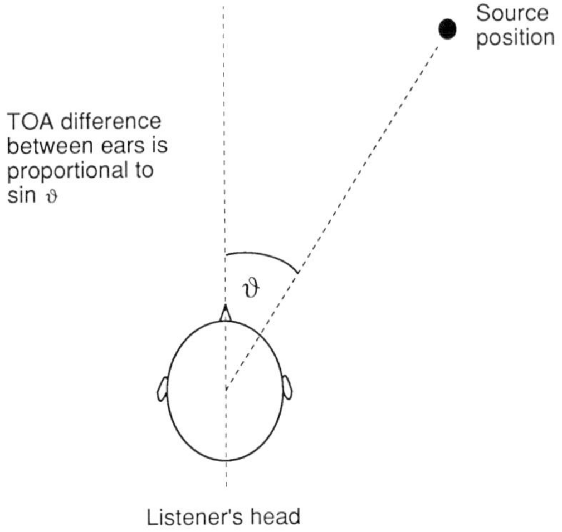

Figure 1.1 Time-of-arrival difference between ears versus angle of incidence of source

At 90° off-centre the time difference amounts to around 600 μs, and it has been shown that the ear is able to detect changes in position of a source corresponding to differences as small as 10 μs.

It should be noted that we rarely keep still when listening to sounds, and even small movements will result in changes of time-of arrival of sounds at the ears. The brain is able to take data from the head's movement and the resulting changes in time-of-arrival to gain more information about the location of the sound. It can be shown that a person whose head is held perfectly steady finds it much more difficult to locate sounds.

Combining level and time cues

Although level cues are important, it can be shown that time cues are probably more important, although one can be traded off against the other to some extent. A moving source of continuous low-to-mid frequency sound is difficult to locate, and even a high frequency continuous sound is not easy. Once the sound is turned regularly off and on it becomes much more easy to locate, even if a hand is placed over one ear (introducing considerable 'bogus' attenuation). It has been shown that if a person wears headphones, and each ear is fed with a short click sound, the clicks will appear to sound as one click even if one ear's click is delayed in relation to the other's. The important point is that the position of the click from left to right depends on the amount of delay, being always towards the advanced ear. With no delay the click appears to be central.

It is possible to fool the brain, counteracting the effect of time difference between the ears, by raising the sound level at the delayed ear. A sound delayed by 0.5 ms between the ears will appear to have moved almost fully over to the advanced ear, but can be 'pulled back' to the centre by raising the level at the delayed ear by a number of decibels.

The Haas effect

A Dutchman named Haas studied the effect of time difference versus level difference in some detail, and arrived at a relationship between the two. His basic conclusion is used every day in public address systems and theatre sound, as well as being evident subconsciously in normal life, being that if there are two arrivals of a sound source at the ears, being the same sound from different directions, the brain will perceive the two sounds as one sound (provided that the delay is not greater than a certain amount), and that the direction will be perceived as that of the sound which

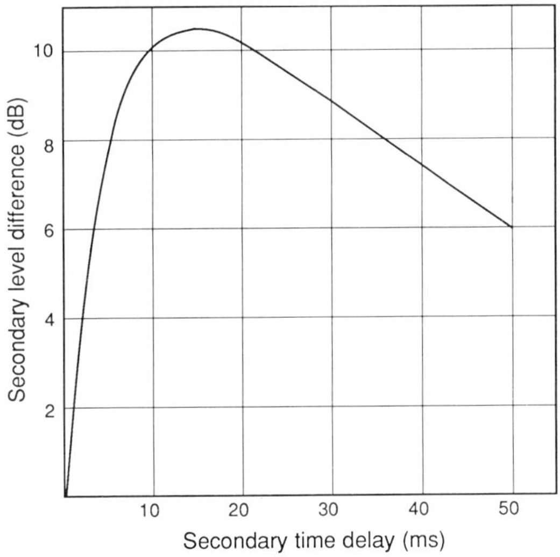

Figure 1.2 The Haas effect—secondary arrival time difference versus secondary level difference for equal apparent loudness

arrives first. More importantly it is even possible for the second source to be louder than the first (within a certain range of delay) before it is perceived as equally loud. The graph in Figure 1.2 shows this relationship, indicating how much louder the second sound has to be for certain delays before it is perceived as equally loud.

Front-to-back and up-down perception

Although less important in two-channel reproduction, the question of perception in planes other than left-to-right will be briefly discussed.

It might be rightly supposed that the time difference for a sound which is so-many degrees off centre will be the same if that sound is in front of the head or behind it, but it is important to realise that a sound which is 40° to the left behind the head will appear to move in a different direction to a sound similarly placed in front of the head if the head is rotated slightly in one direction. Again the ability to move the head is important in locating sounds.

The pinna of the ear has little effect in attenuating sound, except at the highest frequencies, although this may contribute a small amount to front–back distinction. Also, a sound behind cannot be seen, although this may seem an obvious statement, and thus it *must* be behind! Nonetheless, it is still possible to detect the position of sounds with the eyes closed. People have said that when listening to binaural recordings (see below) it is much easier to locate sounds behind them than in front, and this may have something to do with the fact that we use our ears more to locate things behind us where we have no vision. Blind people have claimed to perceive a better frontal sound image with binaural recordings than sighted people, which has some interesting implications.

Memory comes into play in a large way with this area of psycho-acoustics, as our ability to locate sounds is largely to do with the fact that we have learned what things sound like when they are in different positions in a wide variety of different environments. It is the matching of what we hear with a memorized template which is often used to determine direction.

This is very true with up-down perception, as it is usually impossible for a jet aeroplane to scream across underneath one's feet: it is the memory which forces one to accept that the sound must be in a certain place. Listeners who are played two-channel recordings of jet aeroplanes usually duck, even though there is no height information.

Reflections of sound from various parts of the body and surroundings, especially the shoulders and the floor, serve to help in the perception of sounds from certain directions, as the later arrival of the reflected sound results in both a secondary source to be timed, and comb-filtering effects due to phase cancellations at certain frequencies corresponding to the path-length difference involved. The positions of the frequency notches in the perceived spectrum can be compared with memorized patterns of similar situations.

Perception related to sound reproduction

Binaural reproduction

A number of binaural radio broadcasts have been made, as well as other binaural recordings, and these are designed to be played back on headphones, so that each ear only hears one signal channel. To be truly authentic the recording should have been made with microphones in the positions of the listener's ears, with one 'ear signal' recorded on each channel of a tape recorder, but good results have been obtained with 'dummy heads'. It can probably be appreciated that such a recording technique would accurately reproduce the original time-of-arrival and level differences that had been present at the two ears in real life and thus should be very realistic. In practice the degree of reality varies, and, as mentioned above, the front image is not always good although the rear image is usually excellent.

Binaural recording techniques are largely inappropriate in television, as the placing of a real or dummy head is not usually practicable. Also, people rarely listen to television sound on headphones. The results of binaural sound on loudspeakers can be acceptable, but is not usually the best approach for loudspeaker reproduction.

Loudspeaker reproduction

The problem with the concept of loudspeaker reproduction is that each ear will receive the sounds from both loudspeakers. This is unlike headphone reproduction in which each ear receives the sound from only one transducer, and it is thus less easy to imagine how realistic directional sound may be reproduced. Also, there is not now the question of one real sound source to be located at a particular point in space, but of two sources (the loudspeakers) whose job it is to re-create the *impression* of directionality on an imaginary sound stage. Among other achievements, Alan Blumlein, the man famous for developing stereo microphone and disk cutting techniques in the thirties, showed how it was possible to reproduce convincing stereo on loudspeakers by recording not *time* differences between channels, but *level* differences. This is covered further in the chapter on stereo microphone techniques, but the principle to be shown here is that level differences reproduced on loudspeakers actually result in time differences at the ears.

This remarkable statement can be shown to be true with the aid of a diagram. Figure 1.3 shows the situation with steady-state sine waves reproduced on two loudspeakers whose levels differ by a certain amount, but whose phases are the same (no time difference between the sources). This is the situation which would be

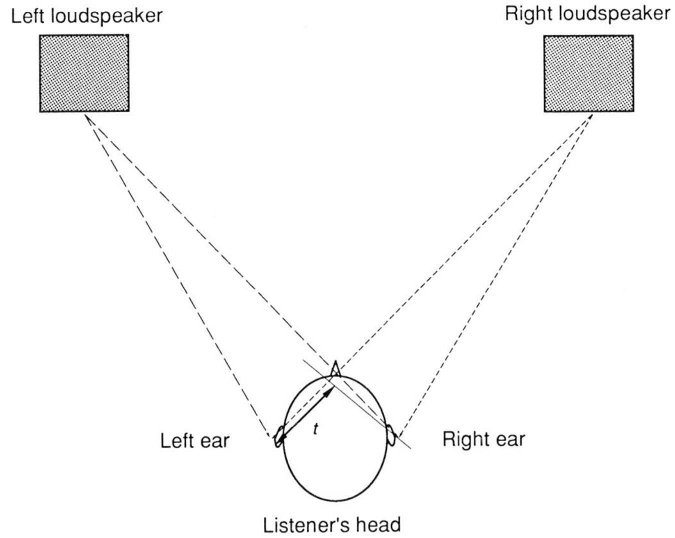

Figure 1.3a Two ears receiving sound from two loudspeakers

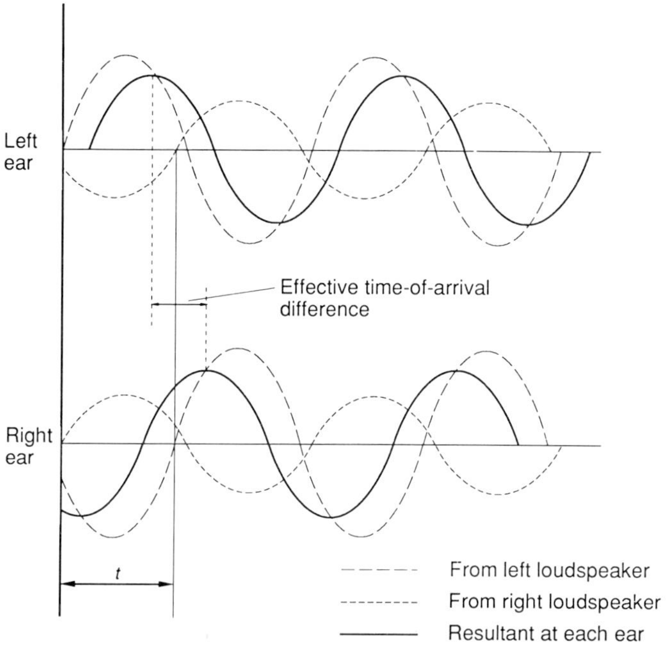

Figure 1.3b Loudspeaker level differences at the ears resulting in apparent time differences. NB: *t* has been exaggerated to illustrate the point more clearly

encountered if a sound source had been panned half left in a mix, say. Sine waves are used for convenience in the diagram, but the same is true for any waveform.

It can be seen that the left loudspeaker signal is higher in level than the right loudspeaker signal, and it will be seen that the lèft ear hears the left loudspeaker first (as it is nearest to it), followed by the right loudspeaker. The converse is true for the right ear, which hears the right loudspeaker first, followed by the left (the actual time differences have been exaggerated somewhat on the diagram for the sake of clarity). If the two sine waves at each ear are summed (which occurs at the ear drum), the result is another sine wave at each ear (the resultant), of different phase and amplitude, indicated on the diagram.

It can be seen that the resultant sine waves at the two ears actually differ in their apparent time-of-arrival, even though no time differences were reproduced at the loudspeakers. So we have created *apparent* TOA differences by the effect of reproducing only level differences on the loudspeakers. The resulting phase difference (or time difference) between the ears depends on the level difference between the speakers, as it may be seen from the diagram that a change in the level of one of the loudspeaker signals will alter the phase of the resulting sine waves.

In recorded sound, both level and phase differences will be used to represent the position of sound sources, depending on the technique used, and it can be shown by experiment that it is possible to trade one off against the other for loudspeaker

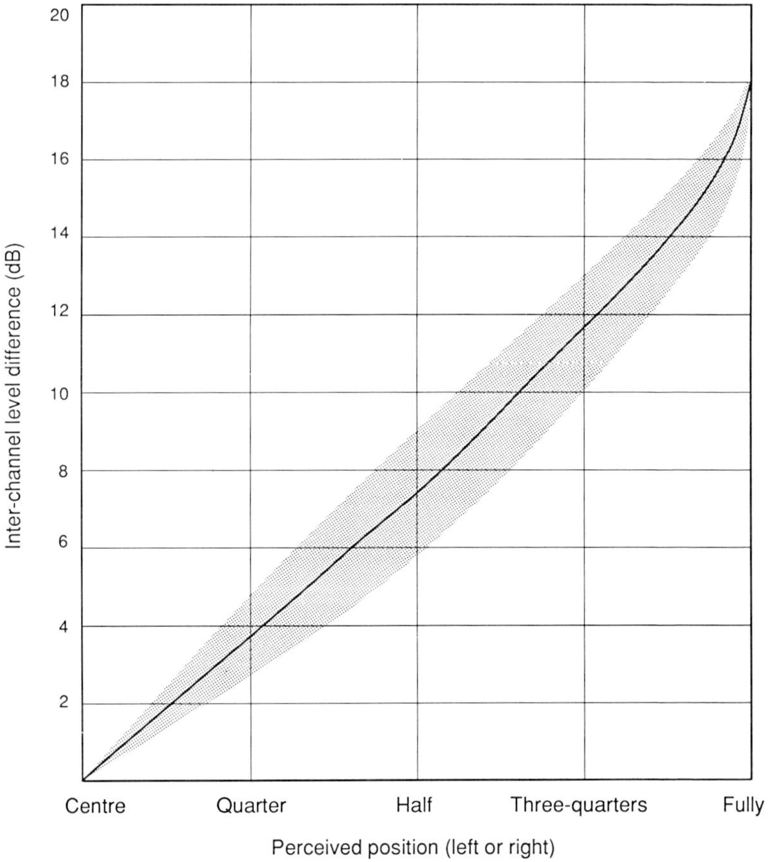

Figure 1.4 Perceived position of a sound source as a result of inter-channel level difference

reproduction. Tests have shown that a level difference between channels of only 18 dB is necessary to give the impression that a sound is either fully left or fully right. The graph in Figure 1.4 shows the perceived position plotted against inter-channel level difference based on a number of test subjects whose perception of position varied according to the shaded area. It can be seen that there is considerable agreement about the level difference required at the extremes and at the centre, but less agreement in between.

It is also possible to show how the effects of level and time difference can be traded-off against each other, by analysing the results from Haas and the experimental results shown in Figure 1.4. The curved lines in Figure 1.5 representing CENTRE, HALF LEFT, and HALF RIGHT, show what combination of time and level difference is required to place a signal in these apparent positions. For example, for a signal to appear in the centre it could have a 2 ms advance on the left channel, but the right channel would have to be 3.5 dB louder to compensate.

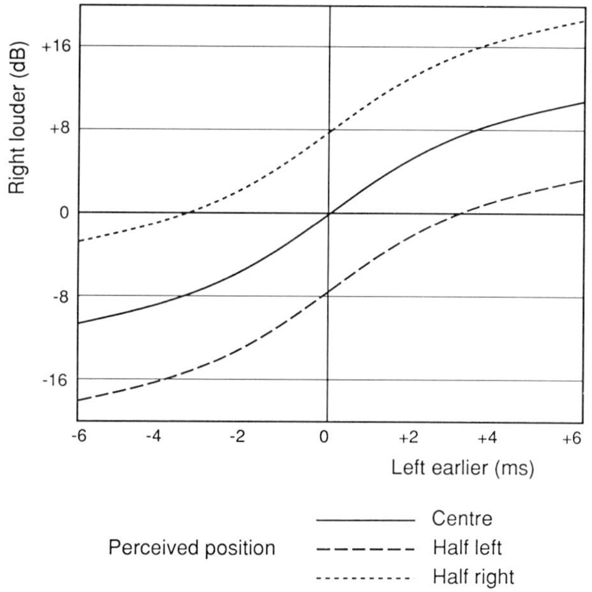

Figure 1.5 Trade-off of level and time differences against perceived position

Implications for broadcast sound signals

Mono compatibility

The theory stated above has important implications in the recording and transmission of stereo signals for broadcasting, as it is desirable not to have phase differences between channels as far as possible, because of the need to maintain mono compatibility. Any phase difference between channels would result in a degree of cancellation if the two channels were summed to mono. Thus the most suitable microphone and mixing techniques are those which result in level differences between the channels to represent stereo positioning, as opposed to those which result in recorded time differences. It would be possible to pan sources artificially using time delay, but it would prove expensive to implement and would result in poor mono compatibility, thus panpots are almost universally based on level difference between channels.

Psychological relationships between vision and stereo sound

Most televisions have relatively small screens when compared with the spacing of loudspeakers for stereo reproduction (see below). This begs the question of what the ideal relationship between the picture size and position and the width of the sound stage is, as one must decide whether it is necessary to hear the sound of a particular source in the picture coming actually from that point in space.

In cinema, the problem does not exist to anything like the same extent, as the picture is roughly the same size as the sound stage and dialogue is largely central, but in television one is dealing with a variety of possibilities for domestic reproduction. It must be accepted that some viewers will have so-called 'stereo televisions' with loudspeakers placed either side of the screen (perhaps detachable), and that some will feed stereo sound from the television through a hi-fi system and larger speakers, probably spaced quite widely. Many may not even place the TV screen centrally between the loudspeakers. Clearly it will not be possible to provide an accurate alignment of visual sources with sound sources which suits all these situations, therefore some compromises must be accepted, and skill exercized, to ensure as wide a range of compatibility as possible. Precisely what is acceptable is very much a matter of personal preference.

There are nonetheless some psycho-acoustic points to be made. One attitude to the problem introduced above could be that however wide the sound stage, the television screen is but a small window on what is really a much larger scene. Most production techniques bear this out, in that a long shot is used to establish the location, then much of the dialogue or action is closer up. Stereo sound might be seen as one way of suggesting off-screen action, or a larger scene than can be viewed, so could be used creatively to advantage. If this is accepted then we have automatically accepted that the picture and the sound do not have to be inextricably aligned and that the brain is capable of accepting sounds in places where there is no picture. Most listeners agree that this is not an unpleasant experience after a short period of acclimatization. It is nonetheless true that an exceptionally wide sound stage is not acceptable with a small screen, as this produces anomalies between the picture and the sound which are extreme.

Positioning of sound sources in a stereo production

Due to the considerable difficulty which can be encountered in ensuring that sound source positions align with their visual sources, many broadcasting organizations recommend that dialogue be monophonic and in the centre, with any effects (either live or dubbed) and music in stereo. This avoids the need for lengthy post-production or stereo microphone technique for dialogue, but maintains the impression of 'stereo space' in the production. The primary rule is that the brain will be confused if there is a direct *contradiction* between what it sees and what it hears: a bus moving from left-to-right in the picture and right-to-left in the sound, for example.

It is not intended to imply that stereo dialogue should be abandoned, indeed it may be a very valuable tool. For it to be perceived as successful considerable planning is required, and careful microphone technique and/or lengthy post-production will be involved. It is also true that dialogue with a *small* amount of width, derived from stereo microphones with the 'S' gain kept low (see Chapter 3), may enhance the impression of space without giving rise to catastrophic problems if mistakes are made.

Relationship between camera shots and sound stage

A final point to be considered is the question of what happens to the sound if the vision changes angle, zooms, or switches to another view. It is most listeners' opinion that abrupt changes in stereo sound image are more disconcerting than abrupt changes in the picture, although this depends considerably on the programme material.

In music, many people find the effect of an 'audible zoom' unpleasant, preferring to see the picture home-in on specific areas of interest, say a soloist in an orchestra, while the sound keeps its original balance. This is probably because most good recording techniques used in the music industry do not involve the fading up of solo orchestral instruments when they play, relying on the composer and the conductor to have ensured that the balance is correct as played by the orchestra. In drama, the opposite may be true, in that a close shot may require an intimate sound from a close microphone, and a longer shot may require a more 'roomy' sound.

In general, it can be taken that if it is possible to leave the position of sound sources in a balance alone then this is largely preferable from the listener's point-of-view, unless a lasting contradiction arises with the picture. For example, if the picture switches to a brief reverse shot of a dialogue scene it is better to leave the sound where it is than to flip the left and right sides to match the picture. The brain has a remarkable ability to accept short-term contradictions between vision and sight when it knows what the real situation is. For this reason it has been suggested that shots which establish the true positions of sources early on in a scene are to be encouraged, whereafter some contradictions may be acceptable.

'Noises Off'

It is much easier for extraneous noises from the production floor to be heard in a high quality stereo balance than it is in mono. Because of the extra directional information, noises such as the dragging of camera cables, talkback, ventilation and other mechanical sources are more easy for the listener to locate and are less well masked by the programme signal. There are thus good reasons for taking more care over extraneous noises in stereo than in mono.

Monitoring and control room acoustics

Control room acoustics

The acoustics of the control room are more important for stereo work than for mono. The principal factor is the question of symmetry, as the stereo image produced at the ears of the balance engineer will depend considerably on his position in relation to the loudspeakers and the surfaces of the room.

Broadcasting organizations have often tolerated fairly poor acoustics in sound control rooms for mono work, but this situation is unlikely to be satisfactory for high quality stereo work. Good acoustic design can transform the sound of a room, often without too much upheaval.

Reverberation time

Little difference is normally expected between the optimum reverberation time for stereo work and that for mono work. The following factors are far more important.

Early reflections

The sound emitted by the loudspeaker will reflect off surfaces in the control room, and this will arrive at the ears of the balancer a certain time after the direct sound. The timing of these reflections can have a great effect on the perceived direction of a sound source, and false information can be given if they are different for the two sides of the head, which would occur in the situation when a room was not symmetrical, perhaps with one side wall further from the balancer than the other. Ideally the room should be symmetrical about the balancing position with early reflections carefully controlled so as not to confuse the stereo image, preferably using a technique such as Time Delay Spectrometry which provides a thorough analysis of energy versus time and frequency at specific positions in the room.

Some early reflections may help to reinforce the volume of the sound, provided that they are within the so-called Haas zone in which the brain integrates subsequent arrivals of a sound source with the first (see above) although reflections with particularly high amplitude are always to be discouraged. Later reflections (that is, later than about 40 ms) will begin to be perceived as discrete reflections, and these will often confuse the stereo image.

Large mixing consoles and other large reflective areas such as equipment bays are perhaps some of the worst culprits in causing high level reflections, but it is not always possible to do anything about these, as they are a vital part of the function of the room. It should be borne in mind that any reflection which results in a path-length difference between the direct sound and the reflected sound to the ears will result in comb-filtering notches in the perceived frequency response at frequencies where the path-length difference amounts to odd multiples of half a wavelength.

Standing waves

Standing waves within a room result in sound pressure peaks and dips in particular locations. This can result in a modification of the perceived frequency response of a loudspeaker depending both on the loudspeaker's position and the listener's position. All rooms have standing waves, and it is a myth that rooms with non-parallel walls discourage their production: making walls non-parallel simply makes the frequencies at which standing waves will occur more difficult to predict.

It is well known that small rooms have worse standing wave problems than large rooms, as there will be a lower number of modes per octave in the low frequency range. This can lead to a 'boomy' bass response which accentuates certain frequencies in one position in the room and cuts them at others, thus the larger the control room the better, as this will increase the number of modes per octave and thus achieve a more even spread of sound pressure as the listener moves his head.

If loudspeakers are placed symmetrically in a symmetrical room then the standing wave pattern should theoretically also be symmetrical, and thus both ears of a centrally-placed listener should experience the same set of conditions. This is the ideal situation, and clearly will not always be achieved, although it will have a positive effect on the balancer's perception of a true stereo image. The point has to be made that the listener at home will be in a far from ideal acoustic position, and that the balance has to suit this listener as much as it suits the television sound engineer. Likewise, the producer and other people in the control room will not be in this ideal 'hot seat'.

Loudspeaker position

Normal monitor loudspeakers

The placing of loudspeakers depends to a large extent on the design of the control room, and the designer of the control room may well have specified particular positions and installed suitable monitoring. If this is not the case then there are some basic guidelines which should be followed:

1. Loudspeakers placed in corners or against walls will suffer bass boost, and may need to be equalised accordingly.
2. The BBC Guide to Acoustic Practice recommends a loudspeaker spacing of between 2 and 2.5 m, although this will depend to some extent on the size of the control room and the size of the TV monitor to which the sound relates (see Figure 1.6).
3. The optimum position for listening is considered to be just outside the apex of an equilateral triangle formed between the listener and the two loudspeakers.

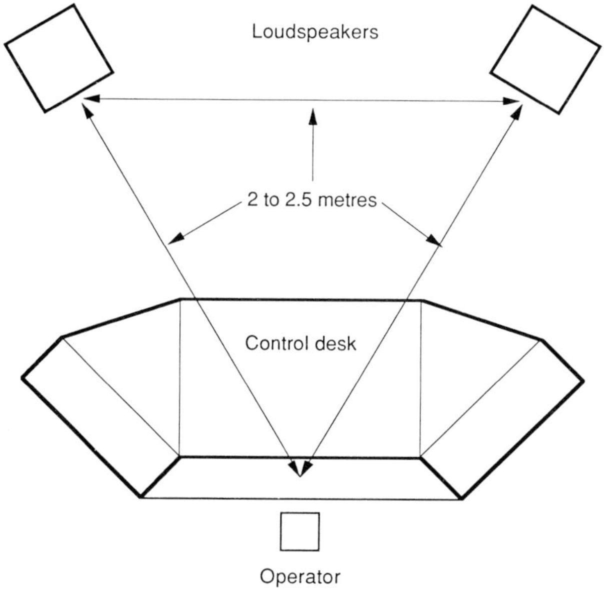

Figure 1.6 Recommended loudspeaker placement for stereo monitoring (Redrawn with permission from BBC Guide to Acoustic Practice)

There is also something to be said for checking the stereo balance on a pair of loudspeakers mounted very close to the sides of a TV monitor, such as would be the case in some domestic situations, although the balancer would have to be at a realistic distance from the monitor to get a proper impression of what the home listener would hear.

Nearfield monitors

Recently, there has been a move towards the use of nearfield monitors in an attempt to combat the problems of poor control room acoustics. Such monitors are smaller loudspeakers which are often mounted on the corners of the mixing console, much closer to the balancer than the normal monitors. This means that they can be at a lower overall level and will predominate over any room effects due to their closeness. Some success has been had in using nearfield monitors in conjunction with a subwoofer under the console to compensate for the lack of low frequency response of the small monitors.

Talkback loudspeakers

If the two main monitors are to be used for stereo programme sound then this precludes the use of one of the monitors for a talkback feed. Clearly a third speaker will be required, and its placement is a matter for personal choice. Some success has been had by a television company which introduced talkback loudspeakers *behind* the heads of the balance engineer and other operators in the sound control room, with the additional directional information aiding the distinction between talkback and programme sound. In this situation it was found that small talkback speakers could be used at a relatively low level, rather like a voice in the ear.

To take this further, it has been suggested that multiple small talkback speakers could be mounted behind the listening position, perhaps suspended from the ceiling, each carrying the sound from a particular person or area; such that the director's voice would emanate from one position, the vision mixer's voice from another, and the producer's from yet another. In this way each talkback speaker would be associated with a person, and the operator would very quickly learn who was who and where.

2

Stereo television system principles

Signal identification

A and B signals

It is conventional in broadcasting to refer to the left channel as the 'A' signal, and the right channel as the 'B' signal. This avoids any confusion when A and B are not the left and right channels of a stereo pair, such as would be the case if they were to carry separate mono languages.

Professional convention holds that the colour coding for these signals is RED for the 'A' signal and GREEN for the 'B' signal (although this may be confusing in the presence of domestic equipment or some European standards where red signifies the right channel). This colouring may be used on metering displays, cabling and controls, among other things.

M and S signals

For mono transmission (which will still accompany stereo sound in most cases) a sum of the A and B signals will be required. This is known as the 'M' signal, for 'main' or 'middle'. This is usually represented by the colour WHITE.

The 'M' signal is not always a simple sum of A and B, as this can result in over-modulation of the mono channel due to the increase in total level that might arise with centrally placed signals, therefore a correction factor is often applied, ranging between − 3 dB and − 6 dB depending on the organization involved.

E.g.: M = (A + B) or (A + B) − 3 dB or (A + B) − 6 dB.

If totally coherent signals existed on A and B channels (representing 'double mono' in effect), then the level of the summed channel (M) would be 6 dB higher than the levels of either of A or B, requiring a correction of − 6 dB in the M channel in order for the mono signal to be reduced a comparable level. In a television production where most of the sound sources are panned centrally (eg: central dialogue) the 6 dB correction factor would be more appropriate. If the A and B signals were non-coherent (random phase relationship), then a 3 dB rise in the level of M would result when A and B were summed, requiring the − 3 dB correction factor to be applied. As most stereo material has a degree of coherence between the channels, especially for central sources, the actual rise in level is likely to be somewhere between the two limits for real programme material.

A suggestion proposed in the United Kingdom, but subsequently rejected, was the introduction of a fixed ninety degree phase shift into the B signal before summing (but not affecting the stereo signal). This would have had the effect of adding a known amount of phase shift between the channels, therefore reducing the rise in level of the M signal when A and B were summed, especially for central signals.

The correction factor must be taken into account when lining up a system (see below).

The 'S' signal (for 'side') results from the subtraction of B from A, with or without the relevant correction, as for the M signal. It is conventionally colour-coded YELLOW.

E.g.: S = (A − B) or (A − B) − 3 dB or (A − B) − 6 dB.

'S' can be used to reconstruct A and B when matrixed in the correct way with the M signal (see below), and it is possible to transmit or record stereo signals in this manner, sometimes known as 'sum and difference'. Although the use of M and S technique in commercial music recording is not great, it is more relevant in broadcasting where the mono signal is still of some considerable importance.

Conversion of signals from A and B format to M and S format

If either A and B or M and S signals are not provided at the output of microphones or mixing equipment it is a relatively simple matter to derive them electrically. Figure 2.1 shows two possible methods.

Figure 2.1(a) shows the use of transformers, where A and B are fed into the primaries; M is derived by wiring two secondaries in series and in phase, while S is derived by wiring two secondaries in series and out-of-phase. Figure 2.1(b) shows the use of summing amplifiers, whereby A and B are summed in-phase to get M, and summed with a phase inversion in the B leg to get S. Both methods have the advantage that M and S may be connected to the inputs of such a matrix, if required, and they will be converted back to A and B.

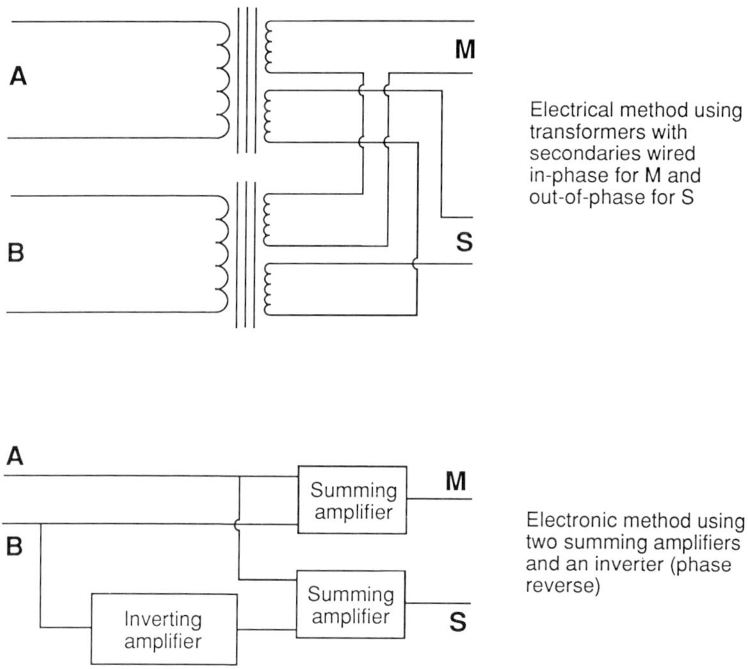

Figure 2.1 Two methods of AB to MS conversion

Phase inversion

It is important that the relative phases of all signals (whether A, B, M or S) are maintained at all times. The effect of a phase inversion in any of these signals will produce potentially disastrous results.

If either A or B is phase inverted, the result will be most unpleasant for the stereo listener, as it is a sonic experience that never arises normally. The stereo image will be imprecise, and central signals will tend to disappear, as will the low frequency content of the signal. The mono listener will hear very little of central signals, as he is now listening to the S signal instead of the M signal. Viz.:
$(A + (-B)) = (A - B) = S$. Conversely, if the S signal is phase reversed then A and B channels swap places (left becomes right, and *vice versa*). This can be useful as a means of correcting for incorrect orientation of an MS microphone (see Chapter 3).

System line-up

Reference levels and line-up tones

Each broadcasting organization works to its own line-up levels, and thus specific rules cannot be given in this area. Also, a number of organizations have yet to finalize their procedure with regard to stereo operation. The following, therefore, are principles to be considered.

As already indicated, if A and B are summed to give M this will normally result in an overall increase in level of M above the peak level of either A or B on its own. The exact rise depends on the degree of coherence between A and B signals. If it is desired that A, B and M should have the same allowable peak level, then the maximum correction factor of − 6 dB should be applied to M in order to account for the 'worst' case when both A and B carry the same signal (say with centrally-panned dialogue). There is a danger in this case that M might be undermodulated with certain types of program material having less central coherent information, leading to a poorer mono signal-to-noise ratio. If this method were chosen, then line-up tones at 0 dBu could be used on A and B, resulting also in 0 dBu on the M signal.

An alternative is to use − 3 dBu tone on A and B for stereo line-up, and for M to be defined as (A + B) − 3 dB. In this scheme, M would read 0 dBu on line-up. The disadvantages are that mono line-up tones (usually at 0 dBu) and stereo line-up tones would be at different levels, and that − 3 dBu is perhaps difficult to read accurately on a peak programme meter (PPM). It is possible, therefore, that some confusion might arise.

This begs the question of which is the correct meter to monitor for stereo transmissions, as correct peak level on M might mean under-modulation of A or B, for example. It should be remembered that a number of stereo transmission systems use digital or companded analogue processes, and thus have a wider dynamic range than the existing mono sound channel. For this reason it might be suggested that correct modulation of the M channel is more important than the risk of slight under-modulation of A or B.

It should be stressed that line-up tones are to be recorded in-phase on A and B.

The setting of absolute audio line-up levels in relation to transmitter modulation levels depends very much on the transmission system in use. In the case of a digital transmission system such as NICAM 728 (see Chapter 6), where pre-emphasis may be used and where limiters might not be used in the interests of high audio quality, the audio reference level must be set low enough such that over-modulation (clipping) is not risked, especially as the pre-emphasis will result in a high frequency boost in the signal before modulation. The EBU (European Broadcasting Union) recommendation in the case of NICAM digital transmissions is for the audio reference level to be set 14.8 dB below 'peak bits' (i.e. maximum modulation) of the digital transmission system at 2 kHz. The overload point of the digital modulator is thus specified as being at + 14.8 dBu.

Identification of line-up tones

In order that line-up tones may be identified as either A or B, it is recommended that some means of differentiation exists between the tones used. An example is the use of interrupted tone on one channel. The most recent CCIR recommendation is to use interrupted tone on A, and continuous tone on B. The interruption should be for 0.25 s every 3 s. It is understood that some

Australian broadcasting companies are recommending a change to the interruption of the B channel instead of the A channel. The BBC has adopted a different means of stereo line-up tone identification which is claimed to be superior as it offers more information to the user. This method involves the use of a single interruption on one channel and two interruptions on the other, such that the mono sum should result in three equally spaced interruptions if everything is set up correctly. If the A and B channels are out-of-phase then the result will be no signal on the mono channel with tiny pips at the interruptions. There is the additional advantage that either channel can be identified as 'stereo-originated' when listened to on its own, whereas in the former scenario the B channel is a continuous tone, and might be confused with a mono recording.

Interchannel differences

Differences in level, frequency response and phase may arise between either A and B or M and S signals at a number of points in the production and distribution chain. It is important that these are kept to a minimum for stereo work, as inter-channel anomalies result in various audible side-effects. Differences will also result in poor mono compatibility. These differences and their effects are discussed below.

Frequency response and level

A difference in level or frequency response between A and B channels will result in a stereo image biased to one side. It is easy to interpret a lack of high frequency (HF) response on one channel as a lack of overall level, and to correct it with increased level rather than equalization. Also, an A channel with excessive HF response compared with that of the B channel will result in the apparent movement of sibilant sounds towards the A loudspeaker.

MS level and response misalignment results in increased crosstalk between the resultant A and B channels, such that if the S level is too low at any frequency the AB signal will become more monophonic, and if it is too high the apparent stereo width will be increased.

Tolerances of interchannel frequency response are usually kept fairly tight by broadcasting organizations: tighter than for mono operation. One example is the IBA (UK) requirement for signal paths to exhibit a maximum of ±0.25 dB deviation at mid frequencies (200 Hz to 10 kHz) between A and B, rising to ±0.5 dB at low and high frequencies.

Phase

Inter-channel phase difference will affect mono compatibility, and it will also affect one's perception of the positioning of sound sources, as described in Chapter 1. Phase differences between A and B channels will result in 'comb-filtering' effects in the derived M signal due to cancellation between the two signals at certain frequencies.

An example of tolerances required for interchannel phase, again from the IBA recommendation for signal paths, shows that a maximum of 15° phase difference is allowed between left and right channels at mid frequencies (200 Hz to 4 kHz), rising gradually to 30° at low and high frequencies.

Crosstalk

It was stated in Chapter 1 that an inter-channel level difference of only 18 dB was required to give the impression of a signal being either fully left or fully right. It might therefore be wondered why it is necessary to have good separation between A and B signals. For stereo work, where A and B are the left and right channels of a pair, the crosstalk requirements need not be as stringent as they would need to be in the situation where A and B are independent mono signals, say in a bi-lingual broadcast. IBA tolerances for AB signal path crosstalk require 50 dB between A and B at mid frequencies (300 Hz to 6 kHz), falling gradually to 35 dB at low frequencies and 40 dB at high frequencies.

Tape machine alignment

The alignment of a tape machine or VTR for stereo is similar to the mono procedure, but there are some points to be considered which do not arise in mono. Digital machines do not usually drift appreciably in frequency response, phase and level alignment, therefore the following notes apply largely to analogue machines.

Tolerances for inter-channel level, phase and frequency response discrepancies on tape machines are stricter for stereo than for mono. The result of misalignment can be more disastrous if the tape recorder is used to record M and S signals instead of A and B. In this case any level anomaly between the tracks will result in apparent crosstalk between the derived A and B channels, which can be quite significant for small degrees of misalignment.

IBA tolerances for interchannel crosstalk and frequency response are shown in the graphs in Figure 2.2. Inter-channel phase can be adjusted on replay by setting the azimuth of the replay head of the tape machine against a reference tape whilst compar-

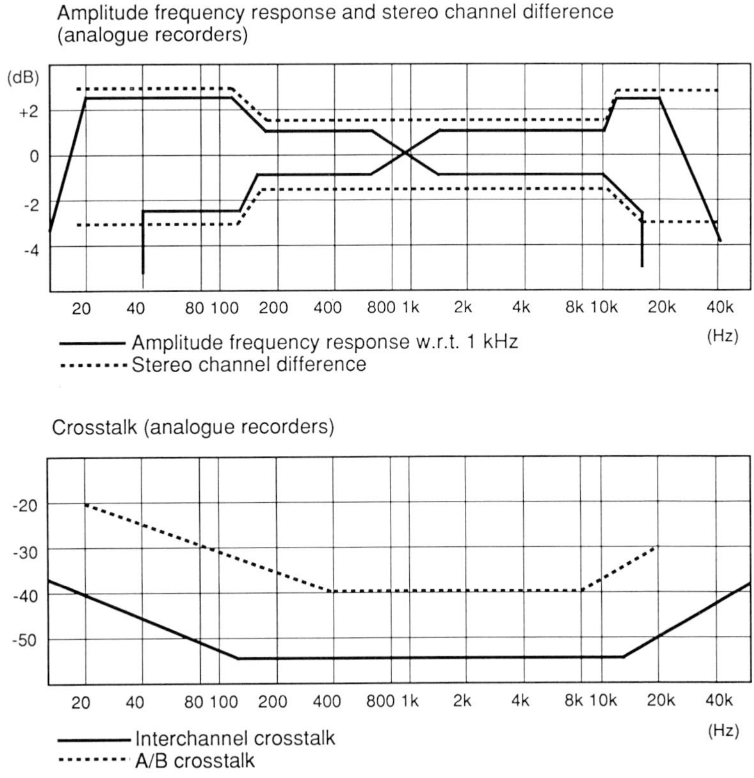

Figure 2.2 IBA inter-channel frequency response and crosstalk tolerances for analogue tape recorders (Courtesy IBA)

ing the relative phases of the two tracks. Ideally a phase meter or scope should be used to check phase differences between channels, but a good visual guide can be obtained by looking at the levels of M and S signals. A lot of information can be gained by viewing a phase meter in conjunction with an MS meter. With a high frequency in-phase tone on both channels, M should be at a maximum when the tracks are phase-aligned correctly, and S should be at a minimum. It may be necessary to add gain to the S meter so as to see the true minimum point (some MS meters have an 'S + 20 dB' switch for this purpose).

It is important to bear in mind that it is possible to align the azimuth wrongly such that the channels are multiples of 360° out-of-phase at high frequencies. At these points the phase difference will still appear to be zero, but the output from tape will not be a maximum. It is recommended that phase should be checked on a low frequency tone to begin with (where the effect on phase of a change in azimuth is not so great), followed by fine adjustment using an HF tone. It is also possible to set phase whilst metering on a stereo vector scope (see Chapter 4), as these may have graticules on the screen which correspond to degrees of phase error.

In setting the record phase it should be remembered that the bias level of each channel will affect inter-channel phase difference, as it affects the apparent recording point on the magnetic tape. Bias should therefore be adjusted *before* setting the record azimuth, rather than after it. The method for setting record head azimuth is similar to that for the replay head, except that a tone is now recorded on both channels and monitored from the (correctly aligned) replay head whilst adjusting the azimuth of the record head using any of the metering devices described above.

Distribution channels

Mono sound signals are routed within television centres often on miles of cable. Distribution of sound over land-lines involves even greater distances. From the points noted above it may be appreciated that considerably greater care should be taken with the distribution of stereo pairs, whether AB or MS, than is necessary with mono signals.

Level, response and phase errors are key features of long

analogue links, and thus it is vital that stereo signals are handled as pairs, as it is only too possible for one leg to pass via an entirely different route to another, perhaps travelling considerably further or being subject to different errors. For this reason the use of digital links is becoming more popular, as a stereo pair can be carried over one digital channel with none of the problems described, except perhaps for the problem of synchronization of a large number of links.

The AES/EBU digital interface is one such link, capable of carrying stereo digital audio at professional rates around studio centres for medium distances (see Chapter 5), and can carry a pair of signals on one ordinary balanced line. Stereo or Dual Sound-in-Syncs is a method which has been developed for distributing two channels of digital sound in the line-sync period of a TV picture, which has the advantage that the sound stays with its picture, and may be used for longer-distance distribution to transmitter sites (see Chapter 5).

3

Stereo microphone techniques

The use of multiple microphones for monophonic work is commonplace in television, as it offers probably the most flexible means of control for the sound mixer. There may be exceptions in certain types of music recording, but it is rare to find productions of any kind using only one microphone. Original stereo recording and reproduction principles pioneered by such people as Blumlein earlier in the century hold that an accurate stereo representation of a sound stage can be captured from two directional microphones and can be reproduced on two loudspeakers. This 'purist' technique is used widely for classical music recording, and may have certain applications in television. Other stereo techniques have arisen, most of which involve the recording of some combination of level difference and time difference from two or more microphones, and it is interesting to note that few of these techniques have any true theoretical basis for their success: the most that can usually be said is that they produce a 'stereo effect'.

It will be appreciated that there are fundamental differences in approach between the use of microphones in television and their use in commercial recording, the most important difference probably being that sound sources and microphones are likely to be moved in television, whereas they tend to remain stationary in commercial recording. In the following chapter, a number of stereo microphone techniques will be examined, and their suitability for television discussed.

The stereo microphone

A stereo microphone is usually made up of two directional microphones in one housing. This is more useful in a practical sense than two separate microphones for the reason that the alignment of the capsules is fixed in at least two planes, and that the capsules can be mounted closer together than two individual microphones, which is important for time-alignment reasons.

Stereo microphones may produce either an AB output or an MS output, or perhaps both. It is also possible that the capsules may be arranged as a pair of either type (see below), with optional matrixing to the other format within the microphone or its control box, and it is usually more practical for microphones of the 'end-fire' type to be of the MS configuration, purely for physical reasons (see Figure 3.1).

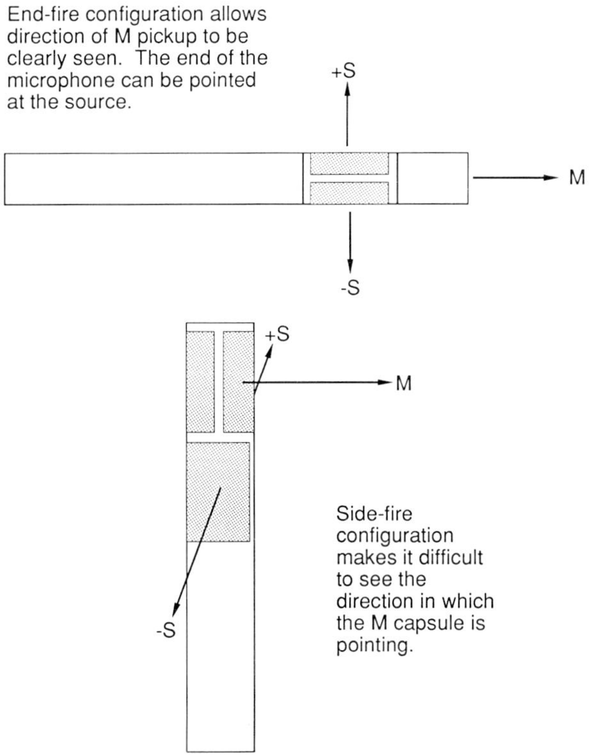

Figure 3.1 'End-fire' and 'side-fire' stereo microphone capsule orientation

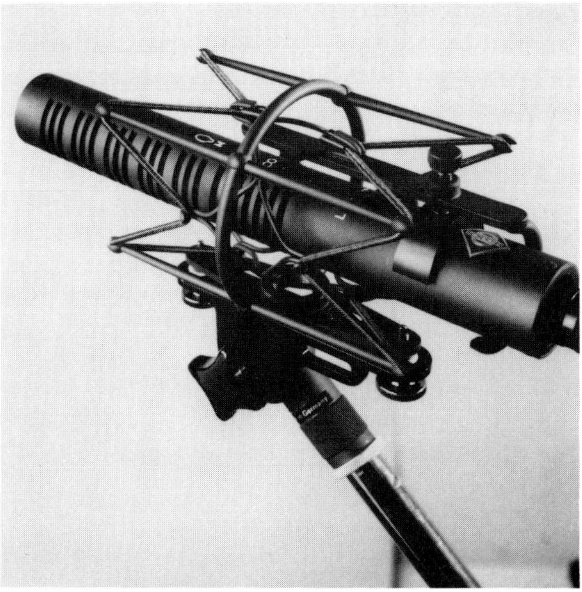

Figure 3.2a Neumann RSM190i—MS stereo microphone (Courtesy: FWO Bauch)

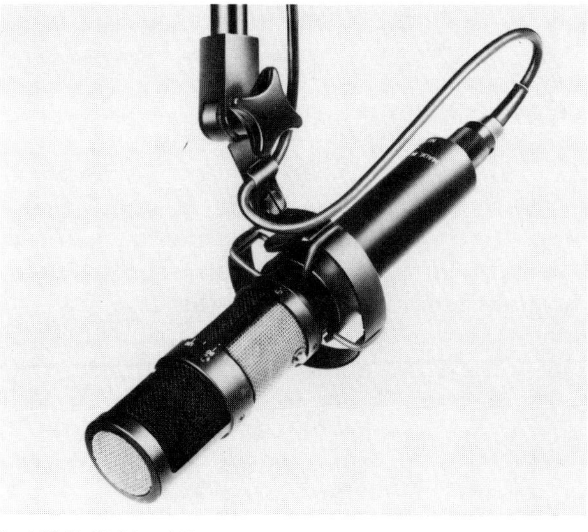

Figure 3.2b AKG C426—AB stereo microphone (Courtesy: AKG)

The photographs in Figure 3.2 show two commercial stereo microphones: the Neumann RSM 190i (an end-fire MS microphone), and the AKG C426B (a side-fire AB microphone).

Co-incident pair principles

Most stereo microphones work on the co-incident pair principle, which involves two capsules with directional pickup pointing in different directions. If the capsules are physically aligned to be at the same point in space as each other (within obvious limits) then there will be no time difference between their outputs because a sound from anywhere within the pickup region will arrive at the two capsules at the same time. This is advantageous when it comes to mono compatibility, for reasons explained in Chapter 1.

Directional information resides solely in the level difference between the outputs of the two capsules, which are a result of their directional pickup pattern. It is interesting to investigate a number of common co-incident pairs, looking particularly at the AB to MS equivalents, as these show what the mono listener will hear when compared with the stereo listener, and throw valuable light on the means of controlling stereo width.

Figure-eight pairs

Two microphones with a figure-eight pickup pattern crossed at 90° provide a useful illustration of the principle of crossed or co-incident pair arrangement (see Figure 3.3).

The AB version appears as two microphones, each pointing 45° off-centre. The frontal pickup of each is positive and the rear lobes are negative (in opposite phase). The MS pair is derived by matrixing the AB signals as described in Chapter 2, the M signal being the straight sum of A and B (in this case) and the S signal being B subtracted from A. Thus the M signal is that which would be gained by pressing the 'mono' button on a mixer's monitoring, and the polar pattern of it can be calculated by adding every point on the A and B polar patterns together through 360°. A number of interesting points arise from an examination of the AB pair in conjunction with its MS equivalent.

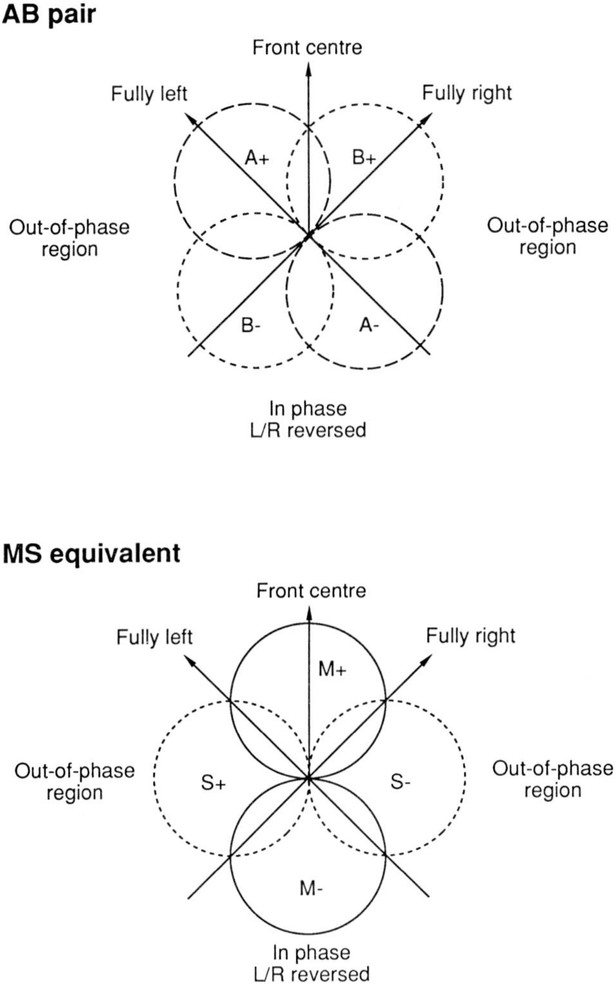

Figure 3.3 Crossed figure-eight microphone pickup pattern showing AB and MS equivalents

Firstly, the MS pair is the AB pair rotated through 45°, thus the mono listener hears a forward-facing figure-eight microphone pickup. Secondly, the sound is fully left where the B capsule's output is at a minimum (this should not be confused with the fact that the A capsule's output is at a maximum at this point: this is simply a coincidence). Thirdly, there is an out-of-phase region around both sides of the AB pair (at an angle of incidence greater than 45°), which can be clearly perceived with signals in this area, and can be verified by listening to a recording of someone walking and talking round such a pair. Fourthly, and most importantly, the out-of-phase region of the AB pair is exactly the same region as that where the S level is greater than the M level.

Specifically, it can be seen that at the point where S = M (at 45° off-centre) A and B start to go out-of-phase, and the sound is either fully left or fully right. From a look at the way in which M and S are derived it can be clearly seen that if either A or B signals are zero then S = M. M = (A + B); S = (A − B). If B = 0, then M = (A + 0) = A, and S = (A − 0) = A also.

This brings out the important point in any stereo operation that if an S level is higher than its corresponding M level then an out-of-phase AB signal is signified. Such a maxim does not apply only to microphones, it is relevant to any stereo pair of signals, and thus can be used as an important guide to the coherence of the signal. The higher S becomes in relation to M, the less coherent the AB signal. If M and S are equal, then the sound is either fully to the left or fully to the right of the AB sound stage.

Another important conclusion is that if the phase of either A or B is reversed, then M becomes S, which could be most annoying for the mono listener. He would then hear no forward information, and a lot of side information.

It would be possible to use a pair of capsules orientated in the MS positions as indicated in Figure 3.3 above, to be matrixed back to A and B before reproduction. This would have the advantage that central signals would be on-axis to the M microphone, thus producing the best frequency response for the mono signal, and that the microphone could be constructed as an 'end-fire' type, with the M capsule pointing out of the end of the microphone, and the S capsule pointing to either side. Thus the microphone could be pointed towards the subject, which has operational advantages (see Figure 3.4).

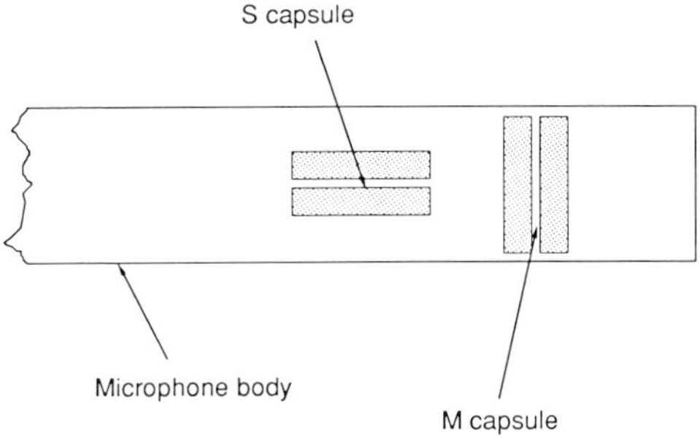

Figure 3.4 'End fire' MS microphone capsule mounting

Other co-incident pairs

The figure-eight polar pattern is not always the most appropriate for stereo television work, as the M signal tends to have too much rear pickup. For this reason the M capsule may have a cardioid response, while the S remains figure-eight. This results in an equivalent AB pair of hypercardioid microphones (see Figure 3.5). Thus the AB pair has less rear pickup, and the mono signal is a forward-facing cardioid, which may be more appropriate for boom or 'fish-pole' work.

One commercial end-fire MS microphone, the Neumann RSM 190i, uses a forward-facing pickup which is close to hypercardioid (it is more directional than the average cardioid). It uses short interference tubes on the forward capsule, and could be said to resemble a 'sawn-off-shotgun microphone' (see Figure 3.2 at the start of this chapter).

The S capsule is a figure-eight. This microphone allows the user to vary the S gain on a control box, in steps of 3 dB, to change the effective width of the sound stage and thus the polar pattern to some extent (see Figure 3.6).

Another commercial example of such an MS microphone is the AKG C34 (a modified version of the original 'AB' C34), in which the polar pattern of the M capsule may be varied between omni and figure-eight (passing through cardioid) so that a boom

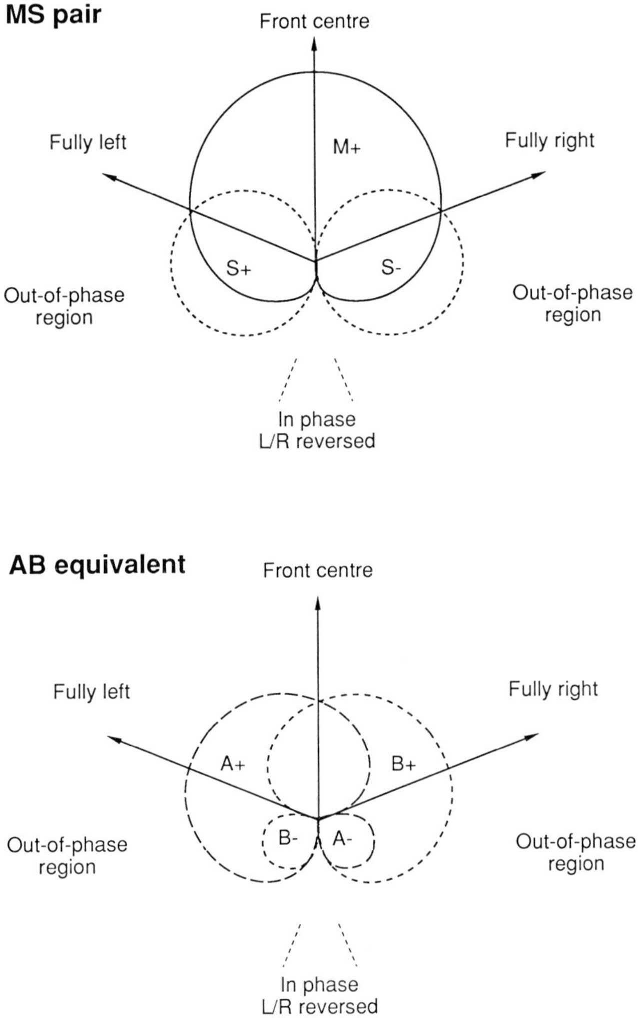

Figure 3.5 MS microphone made up of cardioid and figure eight capsules, showing AB equivalent

operator might widen or narrow the mono pickup to accommodate different situations. The S capsule remains fixed as a figure-eight, and the capsule orientation has been changed from the original C34 side-fire format to an end-fire format.

Varying the stereo width and polar pattern using 'S' gain

By referring to any of the AB and MS equivalent diagrams above it can be seen that if the S gain is varied, the point at which S becomes greater than M will move. Imagine expanding or contracting the S lobes. This has the important result that one can change the point at which 'fully left' or 'fully right' occur in the AB pair by changing the S gain, as we have already concluded that these points occur where S = M.

If the S gain is increased (larger S lobes), the fully left and right points of the AB equivalent move inwards so that they occur at smaller angles of incidence. It is a fallacy that this will produce a narrower sound stage: in fact the narrower the angle between fully left and fully right the wider will be the apparent sound stage, as sound sources which had been say half left will be *further* to the left with a narrower angle. Conversely, if the S gain is reduced the angle becomes greater, as the fully left and right points occur at greater angles of incidence. In this case the reproduced sound stage will become narrower.

S-gain is thus an important tool in determining the degree of width of the stereo sound stage, and may be routed through a fader on the mixing console so that the engineer has control over the width. This in itself can be a good reason for keeping a signal in MS form during the production process, although it can easily be derived at any stage using a conversion matrix. If the number of channels can be afforded on the mixing console, then it would be possible to have, say, one fader for the M signal, one for the S signal, and one for a phase-reversed S signal. In an environment with a number of stereo pair sources, the S-gain of each could be a life-saver for reducing stereo pairs to mono (or very narrow stereo) in times of trouble, such as difficult cross-fades between booms.

The diagram in Figure 3.7 shows how it is possible to derive an AB mix with variable width from an MS microphone using three channels on a mixer. It should be noted that no MS matrices are required. M and S outputs from the microphone are fed in-phase through two mixer channels and faders, and a post-fader feed of S is taken to a third channel line input, being phase-reversed on this channel. The M signal is routed to both left and right mix

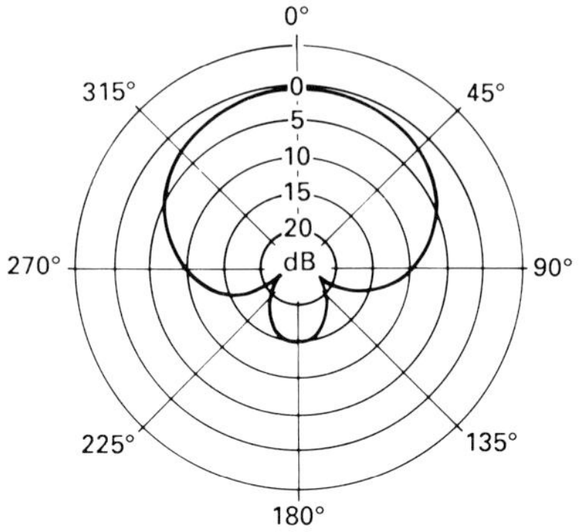
0°
315°
45°
0
5
10
15
20
dB
270°
90°
225°
135°
180°

(a)

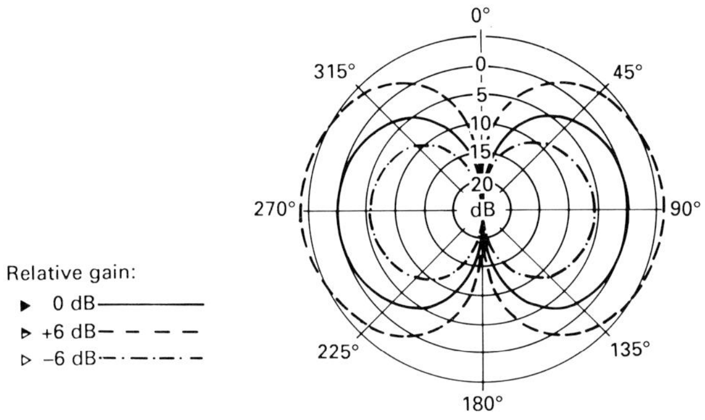
0°
315°
45°
0
5
10
15
20
dB
270°
90°
225°
135°
180°
Relative gain:
0 dB
+6 dB
−6 dB

(b)

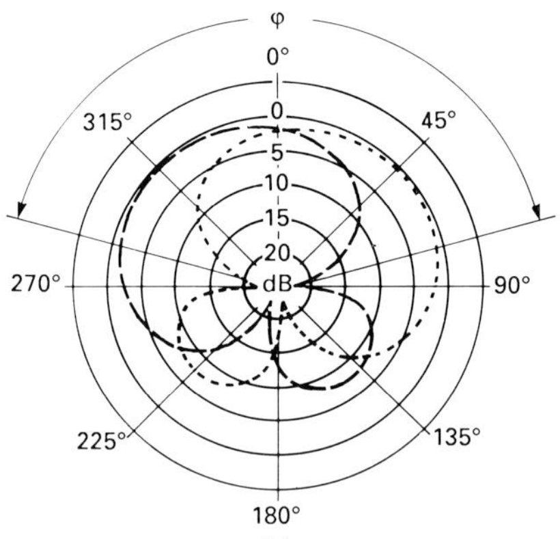
φ
0°
315°
45°
0
5
10
15
20
dB
270°
90°
225°
135°
180°

(c)

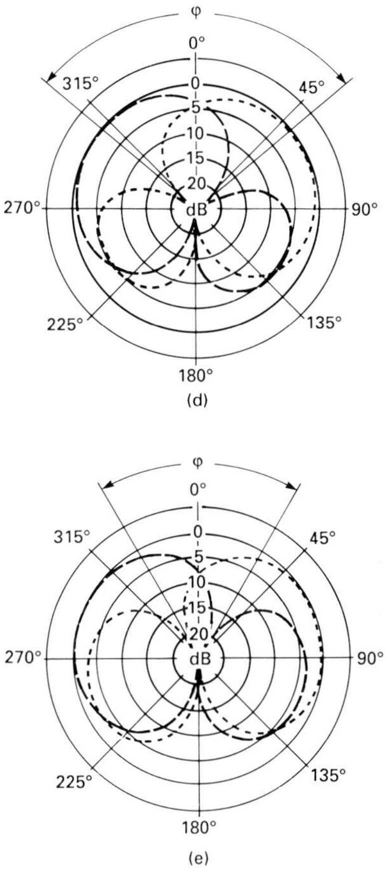

Figure 3.6 Neumann M190i polar patterns. (**a**) M capsule; (**b**) S capsule (variable gain); (**c**) AB equivalent (− 6 dB S gain); (**d**) AB equivalent (0 dB S gain); (**e**) AB equivalent (+ 6 dB S gain)

buses, whilst the S signal is routed to the left mix bus (M + S = 2A) and the − S signal is routed to the right mix bus (M − S = 2B). It is important that the gain of the − S channel is matched very closely with that of the S channel.

A way of deriving M and S from an AB input is to mix A and B together with the B signal phase-reversed to get S, and without the phase reverse to get M. In order to convert back to A and B a similar method to the above could be employed, but this is wasteful of mixer channels and a matrix might be more convenient.

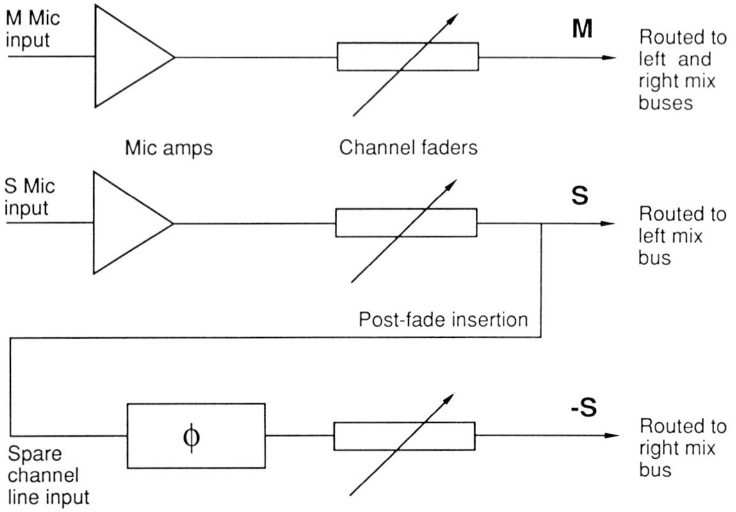

Figure 3.7 A means of deriving a variable-width AB mix from an MS microphone using three mixer channels

A side-effect of changing the S gain is to change the polar pattern of the resultant AB pair. For example, an increase in the S-gain of the cardioid and figure-eight MS pair shown above would increase the size of the rear lobe of the AB hypercardioid equivalent. Thus the audible effect of an increase in S gain is an increase in the rear ambient sound pickup on the stereo signal. It will not affect the mono signal.

Operational considerations with stereo pairs

The degree of usage of stereo microphones may not be great in television work, although initially they may appear to have a number of advantages. Firstly, the adoption of such microphones on booms and poles in TV sets depends very much on whether the TV company in question has decided to opt for central dialogue or not. If central dialogue is decided upon, then many conventional mono techniques will still be used, with stereo microphones only being used to pick up ambience perhaps in live or outdoor environments. Stereo dialogue will require much more planning, and will involve either the use of stereo mics (such as those described above) on booms and poles, or multiple close mics (see below). It may even involve a combination of the two.

Experimental balances have shown that the use of stereo pairs for dialogue can result in an over-wide sound-stage unless the S-gain is kept low, and there is some disagreement on how much width is tolerable in relation to a small picture, especially as one does not know the spacing of the eventual loudspeakers in the home. Subjective tests by a considerable number of experienced listeners seem to indicate a preference for narrow dialogue, with many listeners preferring no width at all. A small amount of width on dialogue can enhance the impression of space in a balance without contradicting the picture, and this is probably best provided by the use of end-fire MS mics used much as normal for mono work but with low-level control of the S signal to spread the resulting AB balance a little.

A problem arises in cases where more than one stereo mic is covering the same scene on a set (in a drama or sit-com for example), as this will require careful cross-fading between booms as sources move across the pickup area. It has been suggested by some operators that a reduction in the S gain of the boom which is being faded down, while fading up the next boom aids a smooth transition.

Outdoors, stereo microphones will be more susceptible to wind noise and rumble than most mono microphones, as they often incorporate velocity-sensitive figure-eight capsules which always give more problems in this respect than omnis or cardioids. Most of the interference would reside in the S channel, and thus would not be a problem to the mono listener.

It would not be wise to use a stereo microphone in the same way as some reporters use a mono microphone when interviewing on the street. The practice of pointing the microphone at whoever is speaking would cause the ambient stereo image to lurch uncomfortably while the subject remained central. Better would be to hold the microphone centrally between the two people and allow the stereo nature of the microphone to pickup the interviewer and his interviewee in their respective positions. In the same way, a stereo microphone on a moving camera would be impractical, as the resulting image would move about all over the place.

Stereo microphones should not be used close to people speaking, as small movements of their heads can cause large changes in the angle of incidence, and thus considerable movement in their apparent position in the sound stage. This is especially true in cases where the S-gain is high, as very small head movements result in swings of the sound from left to right. There may be some conflict between the need for a boom operator to get in close to a sound source for clarity of pickup and the need to keep a certain distance for the reasons stated above.

A final point to bear in mind is the left–right orientation of a stereo microphone. It should be remembered that the rear of a stereo pair (if there is any pickup) will be left–right reversed when compared with the front, and thus moving sounds being picked up on the rear axis will move in the opposite direction to that if they had been picked up on the front. The same occurs if an end-fire MS microphone is held 'upside-down' (that is with the S-capsule facing the opposite way, or phase reversed) because left then becomes right. For this reason most mics of this type have markings on them to indicate the top, and which is left and right! Such a situation of left–right reversal might occur where the rear of a stereo mic was picking up a bus passing which was also in camera vision. The resulting contradiction between sound and picture would be confusing to say the least.

Similarly, a stereo mic on a pole being held 'up-and-under' to pick up dialogue would be in danger of being left–right reversed when compared with a mic being held over the top. It should be noted that this unfortunate situation can be easily rectified electronically by phase-reversing the S signal, and this makes a good case for an S phase-reverse switch somewhere in the chain, as well as an S gain control.

Foldback to boom operators

Success has been reported with the adoption of stereo foldback to boom operators, such that the operator gets a better impression of how well his microphone is placed in relation to the overall sound image. In such an arrangement, left and right signals are fed to the relevant headphones, with talkback fed centrally (equally to left and right). It has even been suggested that talkback be connected to both ears in mono and out-of-phase, so that it appears to be obviously different from programme sound, and placed somewhere indeterminate outside the stereo image.

The Soundfield microphone in television

A variation on the stereo microphone principle is employed in the so-called 'Soundfield' microphone, which can be used in conjunction with its control box to produce either stereo or 'Ambisonic' surround-sound (see Figure 3.8). Although ambisonic sound is not a key feature of stereo TV work, the microphone has found considerable use as a stereo mic because of its somewhat rare ability to be electronically 'steered'. Without physically moving the microphone its polar pattern can be changed, its angle of acceptance changed, and its orientation in all planes altered. This is due to the arrangement of four capsules arranged in a tetrahedral fashion to produce an omni-directional signal and three figure-eights (front–back, up–down, and side-to-side) with electronic matrixing. It is also possible to alter the front–back dominance of the response. When time is limited in setting the position of a stereo microphone, such remote control of its performance is valuable, and this microphone has found use in a number of location recordings, particularly of music in large buildings such as cathedrals, where spot microphones may be difficult to place.

Recently, the same company which produced the Soundfield microphone has employed similar technology in an MS microphone whose orientation may be electronically changed, such that it may be used 'end-fire', 'side-fire' or upside-down.

Omni-directional microphones

The use of omni microphones is widespread in commercial classical music recording, due partially to the smooth and extended frequency response which is not so much a feature of directional microphones. It is possible to record in stereo using spaced omni mics, and the result will be stereo based on both level and time differences between channels. The fact that there are time differences can result in poorer mono compatibility than for a co-incident pair, and this may show up as notches in the derived mono frequency response due to comb-filtering effects. Nonetheless, subjective tests do not always bear this out, with results that

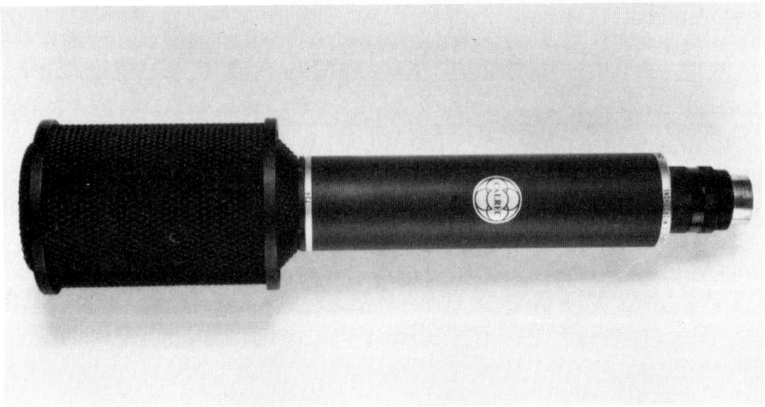

(a)

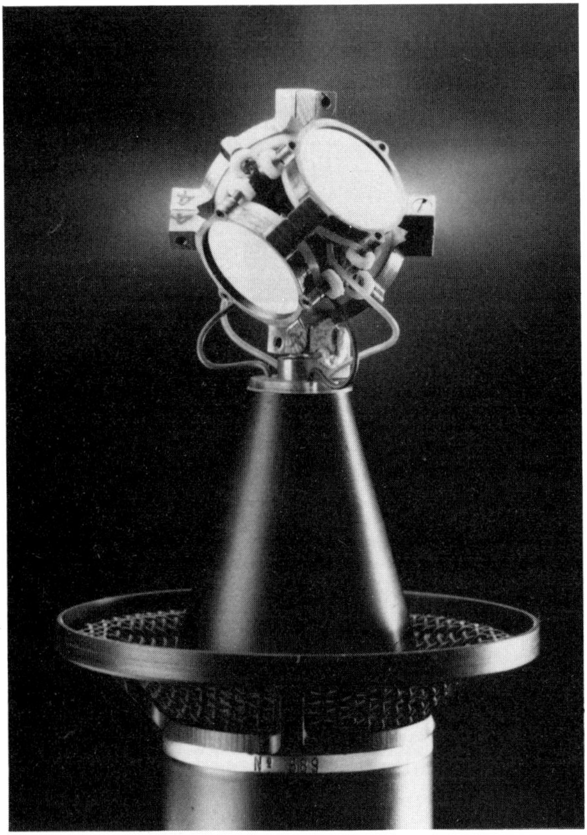

(b)

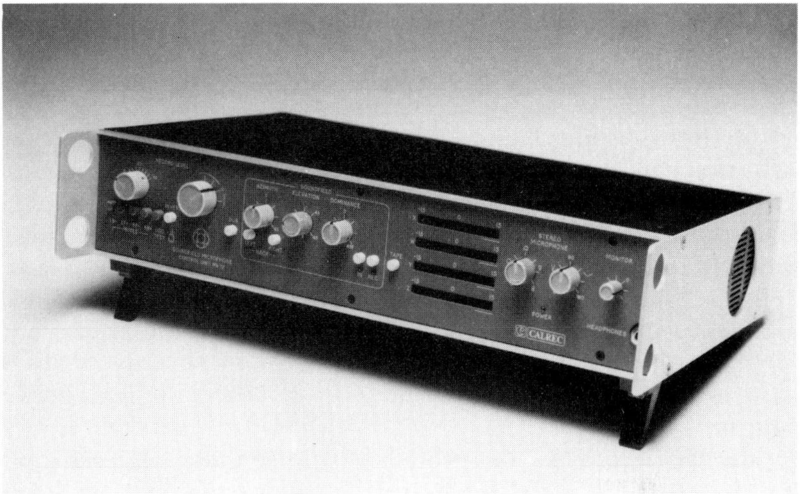

(c)

Figure 3.8 Calrec Soundfield microphone (Courtesy: AMS/Calrec) (**a**) Enclosed; (**b**) Showing capsule arrangement; (**c**) Control box

are often quite convincing in mono. It should be pointed out that a widely spaced group of omni microphones is really acting almost as three spot mics.

There are a number of ways in which omnis may be configured for stereo recording. Two microphones may be spaced apart by between a half and a third the width of the sound stage, or three microphones may be used with the third microphone placed in the centre and slightly forward. The latter has become known as the 'Decca Tree' arrangement, as it is a configuration widely favoured by this recording company (see Figure 3.9).

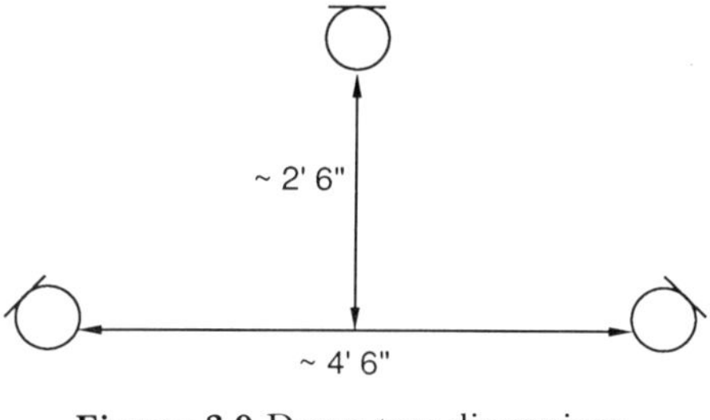

Figure 3.9 Decca tree dimensions

Multi-microphone techniques

Most television sound balancers will already be familiar with the use of multiple microphones for mono work, and it has often been said that stereo multi-mic balancing is easier than mono balancing. There is an element of truth in this statement in that the correct balancing of multiple sources when there is no directional information can be more difficult than when there is directional information. This is because the brain has extra cues by which to identify a sound source, being able to lock onto the *position* of a source as well as its timbre and level. In mono the only means of determining one sound in the presence of others are its loudness and its timbre.

For this reason, anyone who can balance mono material from multiple sources should also be able to balance in stereo, the stereo balance being a matter of using pan-pots to position a source somewhere between left and right as well as controlling its level. Normally, once the panned position of a source has been set, it will be unnecessary to change it, unless a complicated job with considerable post-production is envisaged. If central dialogue is used in a drama, for example, then all dialogue will be panned to the middle, and effects and music may have been originated in stereo anyway, so there will be little to worry about. If stereo dialogue is to be incorporated then some of it may be from stereo booms, while the rest will probably be from close mics, perhaps radio linked. In the ideal situation all sources would be layed to multi-track tape and post-dubbed, but there may be many occasions on which this is not possible for economic ,reasons. In such cases close mics would have to be panned live to the relevant positions, which is more prone to operational error but by no means out of the question, as has been proven in practice.

If extra reverberation is required in the multi-microphone mix it should be borne in mind that less reverberation appears to be required from a subjective point-of-view in stereo than in mono. This could be because of the additional ambient information that is present in most stereo mixes, and which resides mainly in the S channel. The mono listener only hears the M channel. When the M signal is made up of a sum of A and B signals, reverberation, which is largely incoherent information, will suffer partial cancellation. Additionally, the masking effect of the main programme signal on the reverberation will be more pronounced in mono. Even so, it has been shown that artificial reverberation *can* be made more compatible between mono and stereo balances, using careful optimisation of the effect. Operators have suggested that an increase in the crosstalk between the two reverberation returns (making the return more monophonic), or experimentation with phase shift between the return channels can have a beneficial effect on compatibility.

4

Equipment considerations for stereo sound

General

Television companies have been anticipating the arrival of stereo sound for some years, and many will have been gradually re-equipping operational areas for stereo. By no means every item of equipment requires replacement, and satisfactory stereo operation may be possible with only minor changes to the system, but for truly flexible working in the new format there are a number of devices which deserve consideration with regard to their suitability in the future.

The items of equipment to be considered are video tape recorders, monitoring, microphones, mixing consoles and metering. Ancillary equipment such as effects devices, audio tape recorders, cart machines and grams may well be stereo-capable already and will require little change from a hardware point-of-view, although a change in operational practice is clearly required.

Distribution and transmission equipment will normally require attention, depending on the particular systems used, and these matters are discussed in later chapters. Monitoring and stereo microphones have already been discussed in detail in earlier chapters, thus no further coverage will be given here.

Video tape recorders

Until recently, the professional video recorder provided a maximum of three audio tracks. In most cases only two audio tracks have been available, and in some formats there may only be one, due to the need for a track to carry timecode. With stereo sound there is a strong case for the incorporation of a greater number of audio tracks, and it would not be unreasonable to suggest that this number should be double that for mono work. Thus if two tracks are considered only moderately satisfactory for mono work, then four tracks would be required in stereo. Naturally, the true track requirement will depend on how the sound tracks of a VTR are allocated to programme material, and clearly, a VTR with two audio channels is capable of recording a stereo mix, but there would be no space on a two-track VTR for separate 'wild tracks', stereo ambience or music. Additionally there is the long-standing problem of the need to lay-off audio tracks to a separate audio recorder at video edit points so as to allow the cross-fading of 'outgoing' audio material to 'incoming' material. With more audio tracks on the VTR, the separate ATR might be unnecessary, as the lay-off could be made to another VTR audio track.

It is most likely that each of the audio tracks will need to have the facility for 'insert' editing, and that the editing function of the audio tracks will need to be independent of the picture. There is also the question of audio quality, as many of the stereo transmission systems which are being adopted are capable of a very wide dynamic range and frequency response, well in excess of that of older VTR sound tracks. Poor VTR sound quality is likely to become more noticeable as people reproduce television sound through high quality loudspeakers, and it must be remembered that the public is now used to the sound of the Compact Disc.

A number of new video recording formats have now risen to the challenge of high quality sound and a greater number of audio tracks, some of them incorporating digital audio while others make use of analogue noise reduction systems. In the following section the most common video recording formats are examined with relation to their suitability for stereo sound.

U-matic

Two U-matic formats exist, the *high-band* and the *low-band*. This refers to the method of video recording (the high-band format having better picture quality) and does not affect the sound, except for the fact that the low-band U-matic does not have a dedicated timecode track and thus requires one to use the second audio track for this purpose. Low-band U-matics have typically been used in the past in post-production dubbing suites, providing a copy of the edited video master with guide audio, against which further audio tracks can be laid on multitrack tape. This is unlikely to be satisfactory for stereo because the low-band U-matic only provides one usable track of audio, whereas a stereo guide track is preferable.

The audio specification for U-matic machines is relatively poor, having a signal-to-noise ratio of slightly less than 50 dB without noise reduction and a frequency response which just meets broadcast specifications, although Sony recently introduced 'U-matic SP' which attempted to rectify some of the problems and extend the life of the format. Specifically this added Dolby C noise reduction, and slightly improved the frequency response. It would be fair to say that U-matic machines are being superseded by more recent formats.

C-format

C-format reel-to-reel video machines, which have been the mainstay of the high quality VTR market for some time, have somewhat more adequate sound tracks than U-matic machines, and various third-parties have manufactured noise reduction cards which can be used in place of the audio electronics cards in some machines to improve the signal-to-noise ratio. Nonetheless, the audio specifications of C-format sound tracks do not match up to those of a digital transmission system: the analogue tracks of Sony's BVH-2800 PS being quoted as having a S/N ratio of >56 dB on tracks 1 and 2 (no noise reduction), and a frequency response of 50 Hz – 15 kHz, $+1.5/-3$ dB, or 200 Hz – 7.5 kHz ± 1 dB.

There are three analogue linear tracks, the third of which is usually used for linear timecode and has a worse S/N ratio than the other two. Again, therefore, stereo operation would be hindered by the low track capacity. Sony introduced the BVH-2800 machine with two independent digital audio tracks in addition to the analogue tracks, giving a total of four independent audio tracks (five minus the timecode track). A machine such as this might prolong the appeal of the C-format in stereo television, but many broadcasters are looking to newer formats for reasons other than stereo audio.

Betacam

The ordinary Betacam format has two independent analogue audio tracks, and these are offered with Dolby C noise reduction. The introduction of Betacam SP to compete with the M-II format (see below) brought with it two further audio tracks of a different type, notably using frequency modulated carriers which are laid down in the same area of the tape as the video signal by a principle known as 'depth multiplexing', in which the lower modulated frequency of the audio information causes it to be recorded further into the oxide layer than the higher frequency picture which will be recorded nearer the surface. The disadvantage of this method is that it prevents the FM audio from being edited independently of the picture, only allowing it to be edited at the same time and in the same place as the picture. One cannot lay back a post-dubbed multitrack mix to the FM tracks of a Betacam SP machine without also wiping the picture, therefore these tracks are only suitable for the recording of original material.

The advantage of FM recording is the increased dynamic range, freedom from wow and flutter, and better frequency response that is achieved: the performance being almost as good as that of some digital recording channels, with a dynamic range of around 80 dB. If metal particle tape is used with Betacam SP, instead of oxide tape, the S/N ratio of the ordinary analogue tracks is improved from 50 dB to 58 dB without noise reduction.

Initially, the four audio tracks of the Betacam SP format appear to be adequate for stereo operation, but the lack of flexibility in the high quality FM tracks prevents layback and independent editing.

M-II format

The M-II format incorporates FM audio tracks, in a similar way to Betacam SP (described above). Again, these are always linked with the video and cannot be edited separately, thus the same restrictions apply as above. M-II also gives two independent linear audio channels which may be treated with noise reduction.

A digital audio option has been implemented by Panasonic, and this adds two additional channels of professional quality digital audio to the existing complement, with AES/EBU digital interfaces (see below) for input and output. Unfortunately, the space required on the tape for digital audio (recorded at the end of the helical scan which includes the picture) results in the over-writing of linear audio channel 1, and thus removes it from use during digital audio operation. This results in a total of three independently editable audio tracks (two digital and one analogue), with two FM tracks.

Digital video formats

The D-1 (component) and D-2 (composite) digital video formats both offer four independent digital audio tracks. These may be assemble- or insert-edited, and treated in much the same way as if they had been analogue tracks. The advantages of such a format are that the quality of each of the audio tracks is the same, and that there is truly double the normal number of tracks found on older formats.

The BBC, among other organizations, is hoping that D-2 format tapes may be able to be edited and then passed directly to post-dubbing areas without being copied, as the digital format is likely to be robust enough to cope with multiple passes, and it appears practical for digital cassettes to be handled in dubbing suites. The original stereo sound edited alongside the picture, together with any additional material from the other two audio tracks can then be used directly in the dubbing operation without the need for the time-consuming operation of making both working U-matic copies and lay-offs of original audio material to multi-track.

Audio tape recorders—analogue

It is likely that stereo quarter-inch analogue tape recorders will already be in operation in many centres. Older mono machines may be modified for stereo operation in some cases, but otherwise these would have to be replaced.

Two head configurations exist for stereo operation: these are the NAB two-channel arrangement and the DIN stereo arrangement. It is important not to confuse this with the existence of NAB and DIN (CCIR) equalization characteristics, which are another matter altogether. NAB heads are the most common in television for a number of reasons: firstly, they allow true *two-channel* recording as opposed to stereo recording (two-channel recording means that the two channels can be recorded separately and independently), secondly, they use a wide guard-band between the tracks which allows space for a time-code track, and thirdly, the crosstalk is slightly better than with DIN heads.

The use of centre-timecode on quarter-inch tapes is a point that should be considered if synchronization with a picture or accurate location of the tape is required. The centre-timecode standard, as agreed by the EBU, records a narrow timecode track in the guardband between audio tracks, physically co-incident with the audio (see Figure 4.1). This either requires separate heads and a delay line for the timecode track (see Figure 4.2), or specially-made record and reproduce heads which incorporate the time-code head in the same physical entity as the audio heads, ensuring that the crosstalk between timecode and audio is low enough.

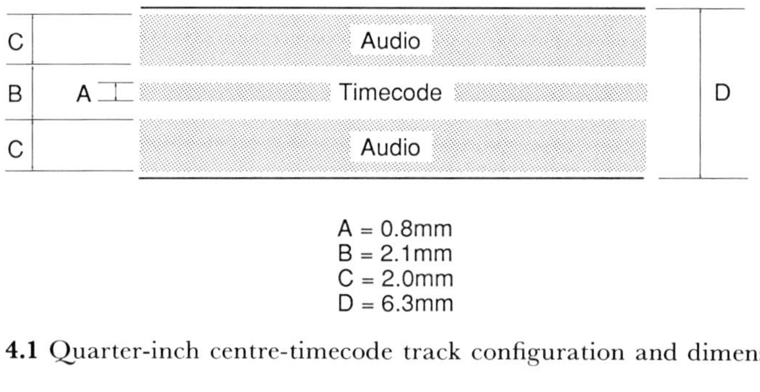

Figure 4.1 Quarter-inch centre-timecode track configuration and dimensions

It must be pointed out that if centre-timecoded tapes are played back on DIN stereo heads there may be some breakthrough of timecode into the audio, as the guardband on DIN heads is narrow. It is not deemed possible to record timecode in this way with DIN heads.

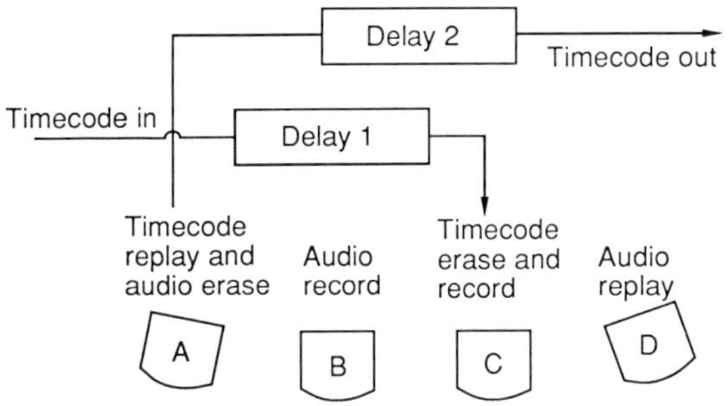

Figure 4.2 Centre-timecode head arrangement using separate timecode heads. Heads A and C are combination heads. Delay 1 is equal to the time taken for the tape to pass from head B to head C, while Delay 2 is equal to the time taken for the tape to pass from head A to head D

Audio tape recorders—digital

The use of digital tape recorders in television will doubtless grow, especially as the gradual move towards an all-digital sound and vision chain takes place. It would be reasonable to suggest that this shift towards digital technology might take perhaps until the end of the century, if not longer, before the whole production and transmission chain is digital. This is especially true of video transmission, which is unlikely to leave the analogue domain for some time.

All the digital audio recording formats that have been introduced to date are capable of handling stereo, but some are not true *two-channel* formats, in that they will not allow independent recording of each channel. This is true of video cassette-based systems, such as the EIAJ format and the Sony PCM-1610 format, as well as R-DAT recorders. Open-reel formats such as DASH and Prodigi, allow independent recording of the two audio channels, as well as providing a timecode track. The DAT format has the capability to handle timecode, either recorded in the subcode part of the helical scan which contains audio as well, or on a very narrow analogue track which resides on the edge of the tape. Not all machines implement this option, and the use of subcode for timecode purposes has not yet been standardized.

The great advantage of digital audio recording, from a point-of-view of stereo television operation, is that it maintains the highest audio quality throughout the recording chain. This is an important factor to consider when the dynamic range of a digital stereo transmission system may be considerable, and its frequency response exemplary. Digital tape recorders rarely drift in response, phase or level alignment in normal operation, and the uniformity between channels is excellent, being well within any broadcasting code-of-practice for stereo. The primary disadvantages at the present time are the cost of digital equipment when compared with analogue technology, and the difficulty of editing encountered in some systems, but it is likely that these problems will become less marked with time.

Many professional systems incorporate AES/EBU digital interfaces, which will interface directly with other digital equipment (see below), but unfortunately the sampling rate of broadcast transmission systems is not the same as the 48 kHz of the professional tape recorder, being 32 kHz in the NICAM 728 system used in the UK. This means that sampling rate conversion would have to be employed prior to transmission, if an all-digital signal chain were to be maintained.

Multi-track tape recorders

Although there is little to be changed about a multi-track recorder for stereo operation, it is possible that a greater number of tracks will be required for post-production work than have been used for mono. Many TV post-production facilities have re-equipped with twenty-four track machines already, although there are still a number of suites using eight-track machines which may need to be updated.

As music recording practice has shown, stereo pairs of channels should, wherever possible, be recorded on adjacent tracks of MTRs, as this will result in the least phase error if there is any azimuth error or instability in the transport.

Stereo mixing equipment

Television companies have for some time been re-equipping operational areas with stereo mixers, and many have replaced older mono mixers with stereo types as a matter of course, even though full stereo operation has been some years away. Without going into mixing console systems in great detail, it would be valuable to introduce some of the options specifically relevant to stereo operation.

Stereo input modules

There may be advantages in being able to handle a stereo source on a single fader, and the stereo input module is an option available on some mixers, taking up no more space than a mono module. Distinction should be made between the stereo *line input* module, which only handles line-level signals such as grams and tape replay, and those modules with stereo *microphone inputs* which may be required in some cases. The facilities on the stereo module may not be quite as comprehensive as those found on the mono version, as there is not the space to duplicate everything, and therefore it is common for such controls as EQ to be ganged between A and B inputs on one set of knobs.

A comprehensive stereo module handling microphone inputs might provide a 'width' control (really a means of adjusting 'S' gain, see Chapter 3), and also a 'balance' control to adjust the relative levels of A and B signals. Auxiliary sends would have to be re-configured to handle stereo signals. There can be a difficulty in metering stereo inputs when multi-track bargraphs for every channel are mounted on a meter bridge, because there simply is not the space for two meters instead of one. This is unlikely to be so much of a problem on broadcast consoles where multitrack metering is less common.

It should be pointed out that so-called 'assignable' mixers make it much more easy to implement stereo channels, as the audio circuits are usually mounted remotely from the control surface. In such a console, one fader may easily be assigned to two audio channels, and the EQ controls may likewise be shared. Systems are available in which the function of physical controls can be changed depending on the mode of the channel (stereo or mono), such that a 'pan' control on a mono channel becomes a 'balance' control in stereo.

Stereo mix buses

Stereo dubbing consoles may provide more than one stereo mix bus, in order to handle the submixes of music, dialogue and effects separately. These may then be mixed down further to a single stereo output, or kept in their sub-mixed form. It is unusual these days to find broadcast consoles with four-channel mix buses, although five-channel buses are not unknown because of the link with film sound operations which can require left, centre, right, front and back buses for surround-sound. It is unlikely that surround-sound productions will be originated to any large degree in television, with companies probably preferring to transmit feature-films with surround-sound encoded into stereo.

Limiters and compressors

Some mixers employ on-board dynamics processing, either on individual channels, or on the main mix buses. For stereo operation it is important that the facility for linking the control of gain between channels exists, so that stereo pairs may be controlled together. This avoids the problems that might otherwise be encountered if say the left channel's dynamics processing were to differ from the right channel's, possibly resulting in a shifting stereo image as the signal content changed. In the case of a limiter, for example, it is possible that an unexpected audio peak in one channel might cause the limiter to act to reduce the level, while the same peak might not be present in the other channel, thus not pushing its limiter into action. This would result in a shift of the image towards the channel whose limiter had not acted, unless the side-chains of the limiters had been ganged, in which case both limiters would duck together, thus maintaining the stability of the image.

A 'link' button is usually provided on dynamics controllers, with an arrow pointing in the direction of the link. This usually only allows linking of adjacent channels' processors, and thus stereo pairs should be kept together on the console.

The method of deriving a mono sum signal from left and right channels to control both limiters is not as effective as the method in which side-chains are linked, as it tends to result in over-compression of central sounds (because the summation effect is greater in the centre).

Interfacing to video editors and switchers

It is becoming more common for high quality audio to be passed through an audio mixer in the video editing suite, as opposed to passing it through the audio switching circuits of video equipment which can sometimes be inferior. This gives the added flexibility of being able to adjust equalization and dynamics, as well as some degree of manual level over-ride.

Remote interfaces exist to a variety of standards and to suit a range of mixers with VCA (voltage-controlled amplifier) faders. Normally, the operator will set up the system so that a specific fader is assigned to the audio from a specific video machine, but for stereo operation it is usually possible to gang the control of certain channels together, although some systems limit this to the simple pairing-up of adjacent channels whereas others allow for free grouping. When considering VTRs with up to four audio channels, flexibility in fader assignment is clearly an important point to be considered.

Stereo metering

General considerations

As the ears cannot be relied on in all cases to give a valid judgement regarding such matters as phase coherence and inter-channel level difference, metering should be taken seriously. It is unlikely, for example, that transmission or master-control engineers will be trained to listen for out-of-phase signals, and they may not monitor in stereo or even have the monitor level turned up very high. Meters provide a quick and easy check for correct connection and alignment. Certain types of metering are suitable for easy visual monitoring of subjective programme levels, and others are more suitable for precise alignment of reference signals. The former may not give such precise numerical information about the number of dBs difference or the number of degrees of phase error, but they may be easier to read in subjective correlation with the sound.

Analogue and digital metering compared

In essence, there are only two types of meter: the analogue type which uses a continuously moving pointer, and the digital type which uses a bargraph or LED display. Broadcast metering in the UK and Europe has typically been of the analogue PPM (peak

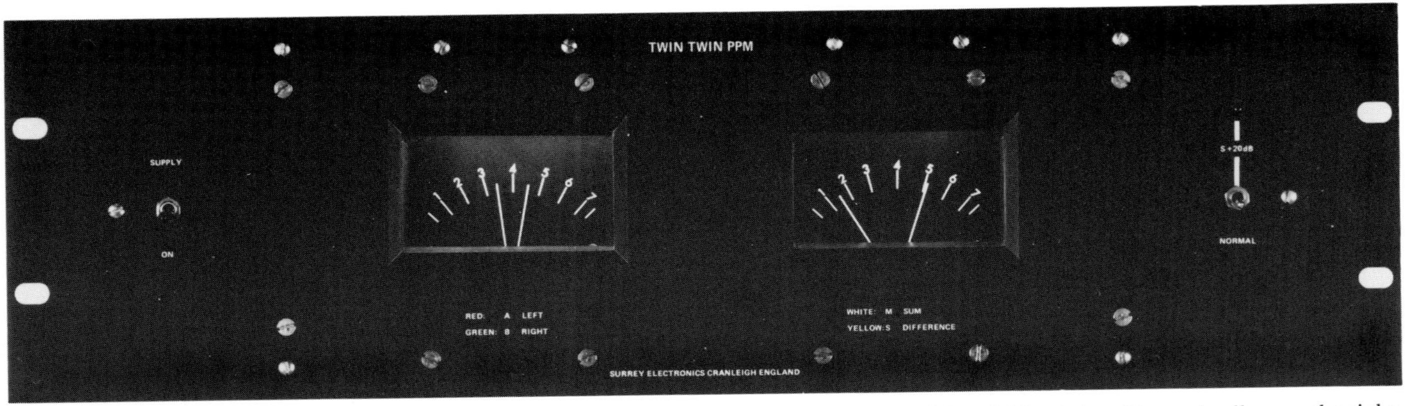

Figure 4.3 Twin stereo PPM showing A and B levels in red and green on the left meter, with M and S levels in white and yellow on the right meter. (Courtesy of Surrey Electronics)

programme meter) type (see below), and in the United States it has been mainly of the analogue VU (volume unit) type. In some organizations these are being superseded by faster digital meters which are capable of reading transients more accurately than moving mechanical meters, although there are many cases in which a reasonably fast mechanical meter, such as the PPM, may provide a better psychological relationship between visual representation and aural perception of peak levels.

European PPMs use a scale measured in decibels, whereas British or BBC-type PPMs use a scale from 1 to 7, with 4 dB between each graticule, except between 1 and 2 where there is 6 dB. The BBC-type PPM thus shows a dynamic range of a little over 26 dB, which is much smaller than that shown on most digital bargraphs. The VU meter has a very slow transient response, rendering it virtually useless for measuring peak levels, although it may be satisfactory for steady-state line-up tones.

Digital metering may be used alongside analogue metering, especially when digital recording or transmission systems are in use, as such systems are especially sensitive to peak overload and will be unlikely to use any limiters to prevent clipping. BBC TV, for example, uses stereo mechanical PPMs alongside digital bargraphs for stereo monitoring.

Stereo PPMs

When metering a stereo pair of signals it is not only the levels of the A and B channels that are important: the relative phase of the pair is also important for line-up, and for determining the degree of coherence between the channels for mono compatibility purposes.

A stereo PPM, incorporating one dual-needle meter for A and B levels, and a second meter for M and S levels, can show most of the necessary information to check both level and phase, as well as being relatively easy to interpret (see Figure 4.3). It also provides an indication of the level of the mono compatible signal (M) and the stereo width (indicated by the level of S). Such a meter shows immediately if the signal is out-of-phase, (S level higher than M level), and indicates the degree of coherence between the A and B legs (less coherence the higher the level of S). It is also possible to see very easily if A and B line-up levels are the same, because the two needles should line up exactly on top of one another, and discrepancies of less than one dB can be judged visually. The use of two separate meters for A and B would make this judgement more difficult.

One problem with stereo PPMs is that in the presence of vision monitors with considerable glare it is sometimes difficult to see ordinary-sized meters, especially in the lowish light levels present in some sound control rooms. It may also be difficult to distinguish between the white and yellow needles of M and S. For sound level metering at any distance, the use of bargraphs which actually emit light is often more suitable, and some of these may even be made to appear in a TV picture.

Bargraph displays of stereo signals

A number of commercial systems have been developed to show A, B, M and S signal levels on digital bargraph displays. Some also incorporate a bargraph for interchannel phase. Devices exist which are capable of displaying these bargraphs onto a TV monitor, either superimposed on a picture or not.

Such displays tend to handle a wider dynamic range than the mechanical meters discussed above, and may be easier to see at a distance or in low light levels than mechanical meters. It is often possible to adjust the peak response and scaling of such meters to conform to a number of laws, such as to imitate a PPM or VU meter, for example.

Vector displays

It can be useful to display a stereo signal on a vector scope, which is similar to the Lissajous-figure display seen on an oscilloscope when X and Y inputs are connected to the left and right channel signals with the time base turned off. Commercial devices exist, such as the one shown in Figure 4.4, which display the left and right information on a CRT marked with fully left and fully right positions, as well as central positions and out-of-phase positions. This particular display also has bargraphs for left and right level, and a third bargraph which can be switched between M, S and an auxiliary input.

Figure 4.5, below, shows the positions of the trace on such a vector display that would be seen in various conditions. This assumes a simple tone source on each channel, as real programme material will show up as a more messy pattern, although it is

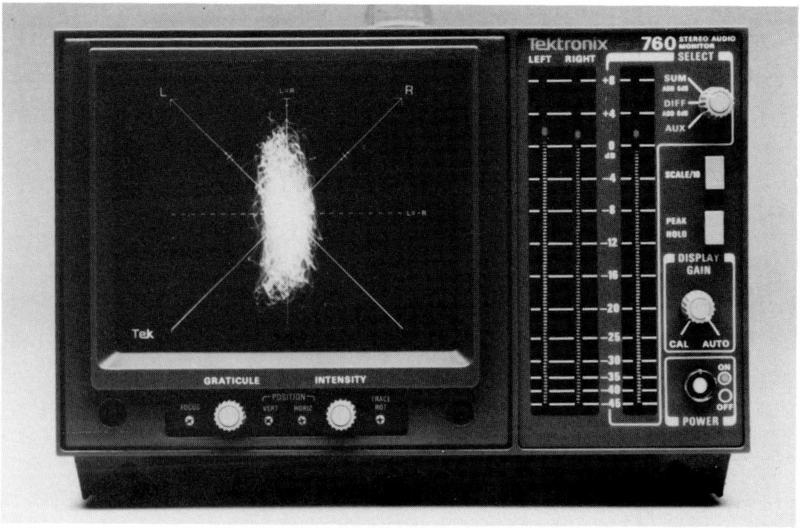

Figure 4.4 Tektronix 760 Stereo Audio Monitor (Courtesy: Tektronix)

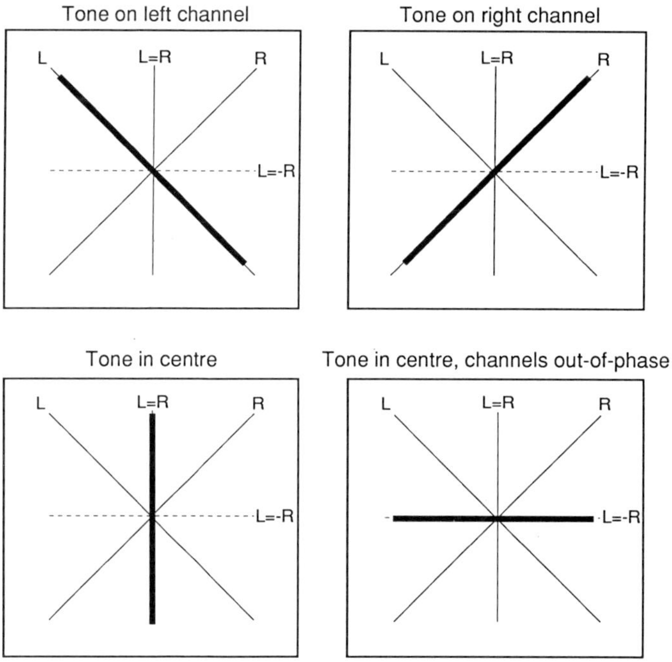

Figure 4.5 Vector scope displays for left, right, centre, and centre out-of-phase tones

surprising how much information can be deduced from the pattern which is produced. A pattern wider than it is high would indicate substantial out-of-phase information, whereas a tall, thin pattern would indicate a more coherent narrow signal. The more circular a central signal appears to be, the greater the phase error between channels.

Phase meters

If one has meters to monitor A, B, M, and S, or a vector scope, it is perhaps unnecessary to have a phase meter as well, but the phase meter can be useful during the line-up of a system where a particular number of degrees difference must be measured between A and B. The phase meter is thus of most use as an engineer's tool.

Phase meters take a number of forms. It is possible to show phase on a mechanical meter with zero degrees in the centre, and plus/minus degrees either side, or on a bargraph with similar scaling, or on a numerical meter which displays the number of degrees phase difference between two signals.

5

Digital signal distribution systems

The distribution of analogue audio signals both within television centres and from the centres to transmitter sites is a complex and specialist operation. Analogue signal distribution within a studio centre is unlikely to have been designed with stereo operation in mind, except in the latest cases, and thus the TV company is faced with the problem of a potential doubling in the number of tie-lines from place to place. This may be impractical and very expensive, considering the miles of cabling that may exist, and thus some companies are looking to a digital means of conquering the problem.

Likewise, a new system is required for the distribution of stereo sound signals with pictures over distances. Analogue land-lines can result in poor coherence of stereo pairs, and existing digital telecommunications circuits must be rented at cost from third parties.

Two digital systems are described below, one of which could be used over medium distances within a studio centre (the AES/EBU interface), and the other of which is designed for carrying stereo sound in the sync period of a video signal (sound-in-syncs) to be used for distribution of signals over larger distances within a TV network or to transmitters.

The AES/EBU digital interface

A primary reason for any television centre to consider the AES/

EBU interface as a means of distributing stereo signals is that it carries two channels of audio over one balanced line, thus existing mono analogue tie-lines could be usefully converted to stereo by the adoption of suitable AES/EBU coders and decoders. The true worth of such an approach will only become apparent when sources and destinations are all in the digital domain, thus requiring no conversion to and from analogue audio, but it may prove more economical, even with analogue audio equipment, to add digital coders and decoders to mono tie lines than to replace all the existing cabling with analogue stereo pairs. There is the added problem with analogue circuits that existing duct space may not allow for the larger or greater number of cables required for stereo.

Data format

The interface involves the serial transmission of stereo digital audio samples of up to 24 bits in length, together with auxiliary data representing both user information and details of the sampling rate, pre-emphasis, copy protection, and so forth. Two audio samples are transferred per audio sample period, and the serial transmission rate is governed by the sampling rate in use (the interface will carry audio at sampling rates between 32 and 48 kHz).

Standard XLR connectors are used at either end of the bus (although presumably other connectors could be used with success), and the electrical interface conforms to RS422 standards. The data block corresponding to two audio samples plus the relevant auxiliary data is known as a frame, and this is divided up into two identical sub-frames (A and B): one per channel. The format of data within the sub-frame is given in Figure 5.1.

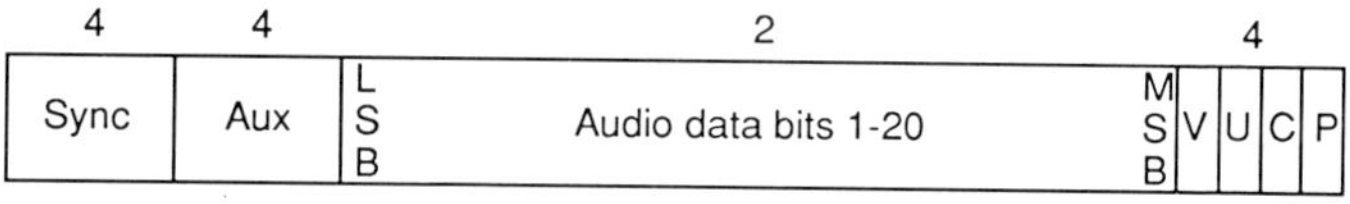

Figure 5.1 AES/EBU sub-frame construction. V = audio sample validity bit, U = user bit, C = channel status bit, P = parity for each subframe

As can be seen from the diagram, the sub-frame is preceded by a pattern of four bits, called the sync block, which identifies the start of the sub-frame, and this is followed by four bits of auxiliary data which may be set aside for additional audio resolution if required. Normally, audio is carried in the twenty bit block which follows (with unused LSBs set to zero if less than twenty bits of resolution are required). Of the V, U, C and P bits which follow, V is a validity bit which indicates whether or not the audio data within the sub-frame is valid or not, U is a user bit (unspecified purpose), C is the channel status bit, and P is a parity bit.

The validity bit is not always used, but there have been cases on some older equipment where a separate AES/EBU interface was provided for each channel of a stereo pair, effectively using it as a mono interface. In this case, every other sub-frame on each interface would be 'blanked' by setting its validity bit to 'false', so that only one channel of audio information would be accepted from each by any receiver. The 'A' subframes of the right channel interface would be set false, and the 'B' subframes of the left channel interface would be set false. It is also possible that the validity bit could be used in large networks where parity errors became more common, acting as a means of signalling to a decoder further down the chain that a particular sample was erroneous.

The channel status bit is used to build up a large digital word over a number of frames. Over 192 frames, a 24 byte channel status word can be transmitted, one bit at a time, and each of the bits of this word can be used to indicate various attributes of the signals being transmitted. For example, byte zero of the channel status word carries the following information:

bit 0:	0 =	consumer use
	1 =	professional use
bit 1:	0 =	normal audio mode
	1 =	non audio mode (eg: data only)
bit 2–4:	000 =	pre-emphasis not indicated, default to no emphasis
	100 =	no emphasis
	110 =	50/15 μs emphasis
	111 =	CCITT J17 emphasis
bit 5:	1 =	source sampling rate unlocked
	0 =	default and source sampling rate locked
bits 6–7:	00 =	sampling frequency not indicated, default to 48 kHz
	01 =	sampling frequency 48 kHz
	10 =	sampling frequency 44.1 kHz
	11 =	sampling frequency 32 kHz

Other uses of the channel status byte are for such things as time information, source and destination identification, and indication of the usage of the aux data block, although these are not always used.

Modulation method

Bi-phase mark coding, essentially the same channel code as used for SMPTE/EBU timecode, is used in order to ensure that the data is self-clocking, of limited bandwidth, DC-free, and polarity independent. A diagram showing the principles of bi-phase mark coding is shown in Figure 5.2.

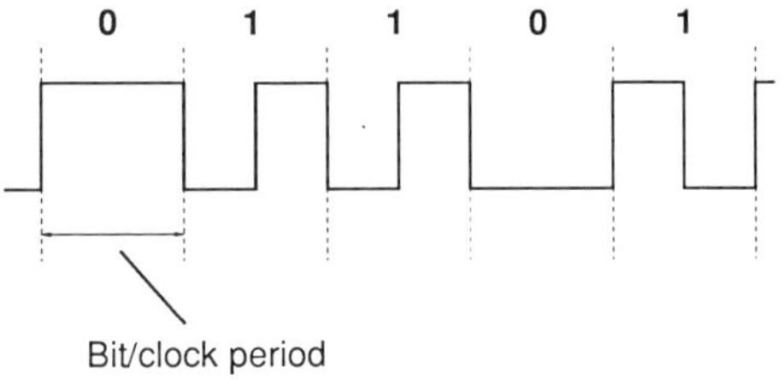

Figure 5.2 Bi-phase mark data coding method

Distance between coders and decoders

The standards document quotes distances of up to 100 m as reliable for AES/EBU transmission over a shielded twisted-pair. It has been found that considerably greater distances than this may be accommodated in many situations, although equalization may be required for longer distances.

Operational considerations

Until such time as audio is maintained entirely in the digital domain, it will be necessary for a number of stages of A/D and D/A conversion to be incorporated in a signal path. As more devices become digital, the number of convertors will be reduced, and the potential for higher signal integrity realized. Nonetheless, it is true that a small number of conversion stages is unlikely to result in a major increase in distortion or noise, especially as most recording devices work to 16 bit resolution, and the final transmission system may only work to 14 bit resolution. Provided that the maximum number of conversions is limited to no more than about three, no serious problem should arise.

Sample rate conversion would have to be employed between the 48 kHz of the AES/EBU interface and the 32 kHz input of a digital transmitter or sound-in-syncs coder (see below), and this will introduce a small amount of delay. Research has shown that this may be reduced to an acceptable level by the use of suitable filters based on the latest DSP (digital signal processing) devices (Bensberg and Emmett, 1988). Signals may also need to be converted back from 32 kHz to 48 kHz, when sound-in-syncs feeds are taken from a large network of television stations, in order to control them on normal studio equipment before being sent out again. The digital audio will need to be pre-emphasized to CCITT J17 recommendation in the digital domain before transmission on a NICAM 728 carrier (see below).

Digital audio signals lend themselves to easy routing without signal degradation, and this is another area in which greater flexibility might be encountered in the use of the AES interface. It is possible for a digital routing matrix to be smaller than its analogue equivalent, and for a number of matrices to be cascaded without reducing the quality of the audio signals.

Dual sound-in-syncs

Developed by the BBC in 1984, and based on some of the principles of an existing system used to carry mono sound, dual sound-in-syncs is a method of accommodating digital sound data within the line sync pulses of a video waveform. Using this method, sound and picture may be carried over a single vision link with no loss of sound quality. The method has the advantage that the digital compansion system used for dual sound-in-syncs is the same as that used to for the NICAM 728 digital stereo transmission system, and thus the bitstream received by the transmitter can be used to modulate a NICAM 728 carrier with little or no modification.

In order to understand the principles of NICAM coding, the reader should refer to the next chapter, where it is discussed in detail in relation to the NICAM 728 transmission system.

Background

The mono SIS system carries two ten-bit digitized audio samples in each line sync pulse of the TV picture by modifying the sync pulse so that it contains digital data. The audio signal is compressed in the analogue domain, and is expanded at the decoder under control of a high frequency signal. A memory buffer and inserter are used at the coder to turn essentially continuous audio data into the burst form required for insertion into the video waveform, and this burst data is re-constructed into continuous form again at the decoder.

It was initially hard to see how twice the amount of data (for stereo) could be squeezed into the same format, as this appeared to involve a complete re-structuring of the video sync information. Research showed that it was possible to change the method of digital coding, such that digitally compressed stereo data could be formed into a serial bit-stream, to be divided up into bit-pairs. Four discrete voltage levels (instead of two) could then be used to represent the four possible states of a sequence of pairs of bits (00, 01, 10, 11) within the sync pulse, as shown in the photograph in Figure 5.3.

This technique results in a lower tolerance of the sound-in-syncs system to noise and interference, but the additional infor-

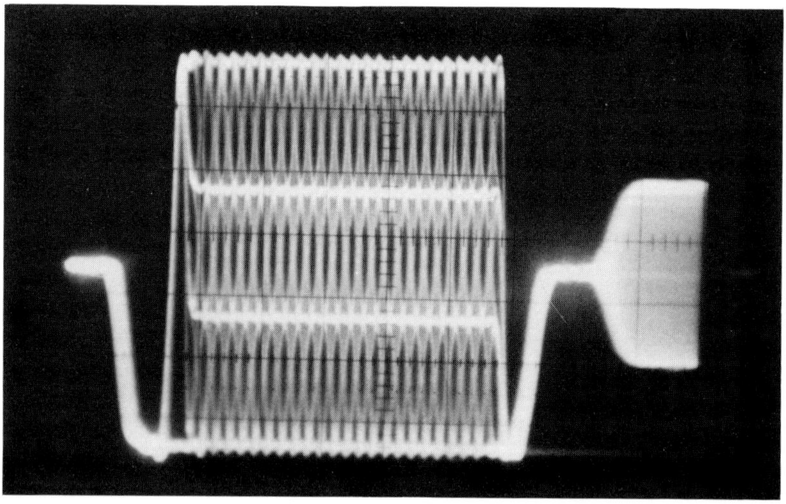

Figure 5.3 Four-level (quaternary) SIS coding within the TV line sync pulse

mation carried proves adequate for stereo operation. It must be remembered that since the vision signal is to be carried over the same link as the sound, the signal-to-noise ratio must be at least good enough to maintain picture quality, and this criterion is considerably in excess of the S/N ratio at which dual SIS fails.

Sampling rate and data rate

Original mono SIS systems operated at an audio sampling rate of 31.25 kHz, the sampling rate being locked to twice the 625 line rate of 15.625 kHz. The NICAM coding method used for dual SIS operates at a sampling rate of 32 kHz, and thus there is no longer a direct relationship between the line rate and the sampling rate. It is important to recognize the difference between NICAM 3- and NICAM 728-based systems, in that the structure of a data frame is not quite the same (see next chapter); also, some dual SIS systems are based inherently on a version of NICAM 3 which results in a data rate of 676 kbit s^{-1}, whereas it has been proposed that future systems may operate at the NICAM 728 transmission rate of 728 kbit s^{-1}.

The BBC's dual SIS system was developed before NICAM 728 was adopted as the EBU standard for digital stereo sound broadcasting with PAL systems I, B and G, although the company was already working on the development of a digital stereo transmission system which has since evolved into NICAM 728. The data rate of 728 kbit s^{-1} was therefore not known as the target. In fact the NICAM 3 system which was used as the basis for the BBC version of dual SIS is not very different from NICAM 728, using the same 14 to 10 bit compression method, and the same sampling rate, therefore audio sample data is compatible between the two systems. The differences lie mainly in the auxiliary control data, the handling of parity information, and the encoding of the scaling factor applied to sample blocks, these being the parameters that result in the different data rates.

Consequently, BBC SIS information must presently be transcoded at the transmitter site from NICAM 3 at 676 kbit s^{-1} to NICAM 728 at 728 kbit s^{-1}, whereas more recent systems, such as that to be adopted by the IBA (UK), are made to carry the full 728 kbit s^{-1} of NICAM 728 on the SIS link and thus involve virtually no treatment of the bitstream at the transmitter. The advantages inherent in the latter approach are that the delay at

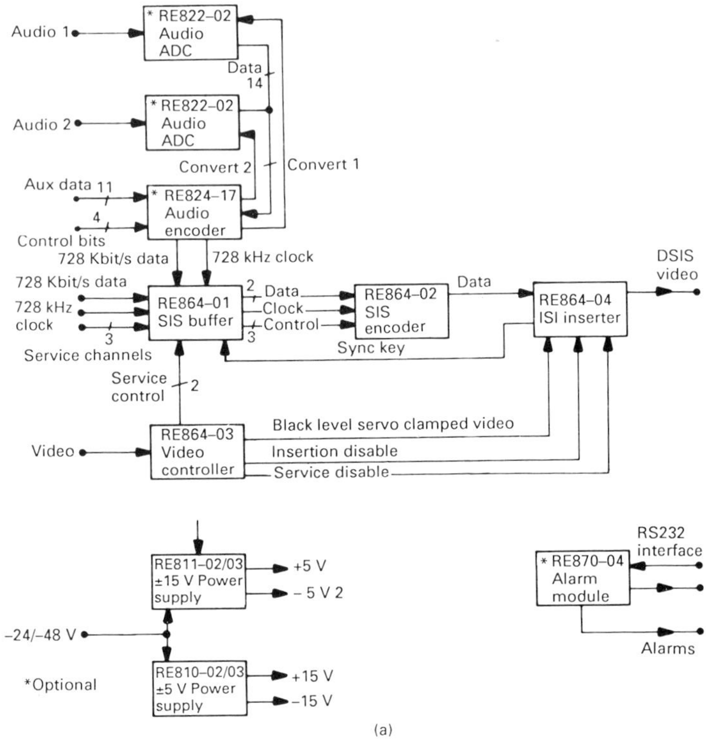

Figure 5.4a Schematic block diagram of the DSIS dual encoder terminal

the transmitter site is minimal, and that any auxiliary or control data could be originated at the studio centre rather than having to be inserted at the transmitter site. It should be pointed out that the four-level coding of bit-pairs in sound-in-syncs corresponds ideally to the four phase states of the 6.552 MHz NICAM carrier which are used to represent bit-pairs for transmission (see next chapter).

Dual SIS has been widely adopted in the UK for the distribution of stereo sound signals around the Independent and BBC television networks, and is used also for distribution of sound to transmitters.

Practical considerations for sound-in-syncs codecs

The block diagrams in Figure 5.4 are taken from a dual sound-in-syncs coder and decoder manufactured by RE Instruments in the

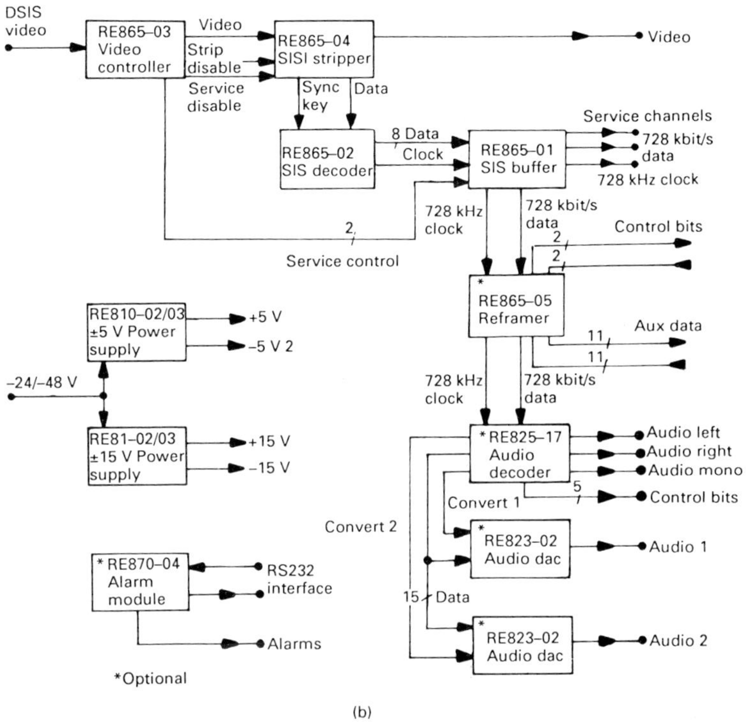

Figure 5.4b Schematic block diagram of the DSIS decoder terminal (Courtesy RE Instruments)

UK. Initially, this system does not have AES/EBU inputs and outputs, but accepts audio in analogue form.

The decoder includes both audio outputs and an output at 728 kbit s^{-1}, together with a 728 kHz clock signal. The scrambling of the data at the 728 kbit s^{-1} output is exactly the same as that required for NICAM 728 transmission (although in fact some additional scrambling has taken place in between coder and decoder), and thus this bitstream may be used directly as an input to the transmitter's DQPSK modulator for the NICAM carrier (see next chapter). In order to provide some additional control at the transmitter site, this data stream will normally be passed through an intermediate device which allows for the interception of any SIS link problems before they are passed on to the listener at home.

The audio section of the RE Instruments coder adds high frequency pre-emphasis according to the recommendations for

NICAM 728, and thus the data output of the decoder will have the correct pre-emphasis for transmission. This raises the interesting question of what will happen when the input to the SIS coder is digital (say from an AES/EBU interface, as above). In this situation, there would be no analogue conversion anywhere between studio centre and the home, requiring that the NICAM 728 pre-emphasis be added in the digital domain. This is not impossible, but would require a re-think of the front-end of the coder, and decisions to be made about the correct point for the addition of pre-emphasis within the distribution chain.

This particular system also incorporates four additional 'service channels', which are not audio channels but signalling channels, carried on four lines of the video waveform within the field blanking period. An alarm signal is carried on one of these, to signal any problems at the encoder to the decoder, and the remaining three are as yet unspecified.

Delays in sound-in-syncs systems

The coding and decoding arrangements inherent in the system described above bring with them an inherent short delay in the audio signal. This amounts to some 7 ms for each codec pair. Provided that the number of codecs is kept to a minimum, this delay should not become subjectively annoying, and it has been pointed out that there may indeed be an improvement over the existing delays encountered in the telecommunications circuits used for mono sound. Psychologically, a delay in the sound which places it slightly after the picture is less disturbing than an advance in the sound, because this is the situation that occurs with natural sound sources in real life! The speed of sound is slower than that of light, and thus one is used to seeing something a short time before one hears it.

6

Transmission systems

The requirements of different countries for multi-channel sound transmission differ considerably, with two audio channels being the obvious minimum for stereo. Many countries have a requirement for two or three languages to be broadcast, especially where there are many dialects (Yugoslavia, for example), and this limits the potential for stereo sound as there is usually a limited bandwidth in which to transmit extra material. The systems used or proposed fall into two basic camps: digital and analogue, and it is only recently that the digital systems have been considered feasible from a commercial point-of-view, whilst analogue systems have been in use in countries like West Germany for some years. Digital systems have the advantage of very high audio quality, with resilience to interference and noise in the transmission channel, whilst analogue systems are also capable of moderately high audio quality and may be cheaper to implement, although with arguably less flexibility and lower tolerance to interference.

A number of clear standards have arisen in key areas of the world. The digital NICAM 728 system for the UK, recommended by the EBU for adoption with PAL television systems in the future; the analogue BTSC or MTS system defined by the Broadcast Television Standards Committee for the USA and Canada; the dual FM carrier system originated in West Germany, and used in some other parts of Europe; the FM/FM system used in Japan; and the Dolby ADM digital system used for some applications in Australia, and under trial in other parts of the world .

In this chapter, the principles of each of the major systems are examined.

NICAM 728

NICAM 728 is the system adopted in the UK for the transmission of two additional high-quality digital sound channels with terrestrial television. It is suitable for use with the PAL System I television standard, and can be adapted for use with PAL Systems B, G and H. NICAM 728 is to be adopted in parts of Scandinavia (PAL-B), and is recommended by the EBU to any member organization which is considering digital multi-channel sound transmission with television.

NICAM 728 provides two channels of sound which can be used for stereo or bi-lingual purposes, in addition to the mono FM carrier which is retained for compatibility with existing receivers. A third mode allows for the transmission of data only, as opposed to audio data, although the audio mode also has a small capacity for data.

Digital audio coding

It is an impractical proposition to transmit digital audio at the same data rate as used in professional digital recording systems or the Compact Disc, due to the limited band-space available between one vision carrier and the next. In PAL-I the inter-carrier spacing is 8 MHz, and the existing FM sound carrier is at 6 MHz above the main vision carrier frequency (see Figure 6.1).

An additional space restriction is caused by the existence of a lower vestigial sideband of the adjacent vision channel, which extends to 1.75 MHz below the vision carrier frequency. The space restriction requires that every effort is made to reduce the bandwidth required by any additional sound carrier, whilst maintaining as high a sound quality as possible, at the same time as ensuring a minimum of interference between the new sound and the picture.

NICAM stands for Near-Instantaneously-Companded Audio Multiplex, and, as its name suggests, involves compansion techniques to compress the dynamic range for transmission and expand it on reception. The compression of the dynamic range of

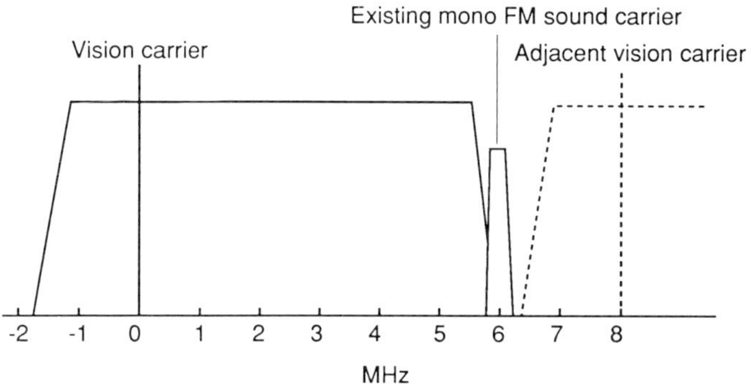

Figure 6.1 Vision carrier spacing in PAL System I, with FM sound carrier position indicated

a digital audio signal has the advantage that a lower number of bits can be used to represent each sample, thus reducing the data rate. Also, the sampling rate used in broadcast communication applications is 32 kHz, as opposed to the 44.1 or 48 kHz of professional audio systems, which further reduces the data rate. 32 kHz has been deemed satisfactory for the purpose of stereo sound transmission, as this gives the maximum audio frequency response of 15 kHz which has been used in high quality FM radio for many years, and is at the upper limit of the majority of the population's hearing. The reduction in data rate that this sampling frequency gives is very large compared with the small loss of high frequency information that it entails.

The 'near instantaneous' part of the title refers to the fact that the dynamic range is not instantaneously compressed, but that it is compressed in groups of 32 samples, lasting 1 ms. The compression method is digital, with no treatment of the dynamic range in the analogue domain, and involves the reduction of the number of bits per sample from 14 to 10, with a scaling word sent for every compressed block so that the samples can be restored to their original levels on decoding. Fourteen bit linear quantization has been chosen as the original resolution before compression, and although this may not be adequate for entirely unpredictable audio signals, it is adequate for signals whose dynamic range has already been controlled (as television audio signals will normally have been).

The system is based on the precept that the higher the level of the audio signal, the more bits it loses in the compression process, and thus the greater the error in quantization. The amount of compression used depends on the level of the highest sample within the 1 ms block of 32 samples. The number of bits per sample is to be reduced to 10, thus sample groups in which the highest-level sample uses no more than the 10 LSB's (least significant bits) will remain un-modified, and thus will be reconstructed perfectly at the decoder. Above this there are then a further four level ranges (for original sample groups using up to 11, 12, 13 and 14 bits) in which samples will lose increasingly more of their less significant bits in order to reduce them to 10 bits in length. A scaling factor code is sent with each block to indicate the original range of the block, and this is used to re-position the samples from a block in the correct range when they are decoded. The missing bits are reconstituted as a digital 'half', in order to minimize the average error which results from the loss of less significant bits.

The audible result of this companion technique is that low level signals (up to 10 bits) are reproduced with no loss of accuracy, but higher level signals will suffer increased programme modulated noise due to a quantization error which becomes greater as more LSBs are lost. Pre-emphasis is applied to the audio signal prior to compression (to CCITT Recommendation J17) with subsequent de-emphasis after the decoder, which further helps to reduce the audible effects of compansion, although at the expense of high frequency headroom. The increased noise at high audio levels is masked by the audio signal, and thus is not so subjectively annoying. Subjective tests have shown that the NICAM coding method is extremely effective, and that there are very few listeners who can detect the difference between a signal which has been companded and the original 14-bit linear signal.

Error correction and range coding

In NICAM 728, a parity bit is added to each 10-bit sample after compression. This parity bit is used to check the six most significant bits of the sample for errors, the parity type being even. Unlike NICAM 3 (an older BBC design based on similar principles), NICAM 728 sends the scaling factor for the audio sample group by modifying the parity bits according to a complex table which dictates how each bit should be modified depending on the position of the sample in a group. This helps to keep the number of extra bits used to a minimum.

Format of data

Once the audio has been sampled, compressed, and subjected to the range-coding and parity process, the data is formed into frames which are 728 bits in length. One of these frames is transmitted every ms, thus the data rate is the 728 kbit s^{-1} implied in the title of the system. One 728 bit frame is made up as follows:

8-bit frame alignment word	= 8
5 bits of control information	= 5
11 bits of additional data	= 11
704 bits of sound, parity or data	= 704
Total	= 728 bits

The *frame alignment word* is a unique pattern of bits that identifies the start of a frame, having the binary form 01001110.

The five *control bits* act as follows:

Bit C0 is the *frame flag bit*, which is set to 1 for eight successive frames, and to 0 for the next eight frames. In this way a 16 frame sequence is defined, which is important because the switching of the mode of operation of the NICAM channel (controlled by bits C1, C2 and C3) occurs at the start of the 16 frame sequence following that in which bits C1, C2 and C3 are changed.

Bits C1,C2 and C3 form a pattern which is used to define the application of the NICAM channel (stereo, mono, bi-lingual, etc). The assignment of these patterns is shown in Table 6.1.

Bit C4 is the *reserve sound switching flag* which is used to indicate that the sound material on the mono FM channel is the same as that on the NICAM channel. This is so that a decoder may tell

Table 6.1 Assignment of NICAM control bit patterns C1, C2, C3

State of control bits			*Application specified (contents of 704-bit sound/data block)*
C1	*C2*	*C3*	
0	0	0	Stereo signal comprising alternate A-channel and B-channel samples
0	1	0	Two independent mono sound signals (designated M1 and M2) transmitted in alternate frames
1	0	0	One mono signal and one 352 kbit s^{-1} transparent data channel transmitted in alternate frames
1	1	0	One 704 kbit s^{-1} transparent data channel

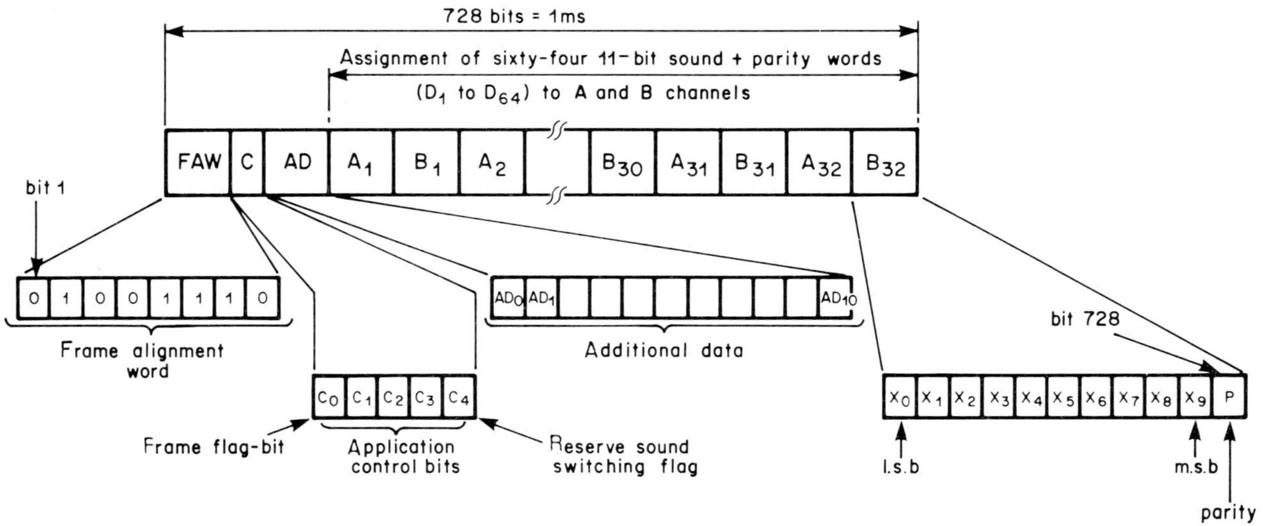

Figure 6.2 Structure of a NICAM 728 frame (diagrammatic) (courtesy BBC)

whether it is reasonable to switch over from NICAM sound to FM sound in the event of a failure of the NICAM channel. Clearly this would only be sensible when the two carriers are handling the same programme material, and bit C5 is set to a 1 in this case. In the case of dual mono or bi-lingual broadcasts, this comparison refers to mono channel 1 only.

The additional data portion of the frame is set aside for the carrying of data at a rate of 11 kbit s^{-1}. The application of these bits is not yet defined.

The use of the remaining 704 bits depends on the application:

Stereo mode

32 samples of stereo audio (= 32 × 10 × 2 or 640 bits) plus one parity bit per sample (= 64 bits), such that the total equals 704 bits, with odd samples carrying the 'A' channel and even samples carrying the 'B' channel.

Dual mono or bi-lingual mode

64 samples of mono audio plus one parity bit per sample, with mono 1 carried in odd-numbered frames and mono 2 in even-numbered frames.

Single mono mode

32 samples of mono audio plus one parity bit per sample, and 352 bits of data, which may be used or not as required, with mono sound being carried in the odd-numbered frames.

Data mode

704 bits of data, the format and function of which is not yet defined.

Transmission

Once in the 728-bit frame-format described above, the 704-bit portion of the data is interleaved (in other words, the order of the samples is changed to place samples which would otherwise have been adjacent at least 16 sample periods apart) in order to minimize the effect of burst errors which would otherwise destroy a contiguous block of samples. The whole frame is then scrambled prior to transmission, with the exception of the frame-alignment word, in order to shape the spectrum of the resulting transmission so as to spread the transmitted energy pseudo-randomly across the bandwidth used. The data must therefore be de-scrambled and de-interleaved at the receiver.

In order to modulate the carrier, which is at 6.552 MHz above the frequency of the vision carrier in PAL System I, the serial data stream formed as above is split up into pairs of bits, and the state of each pair is used to determine the relative phase of the carrier. This is known as DQPSK or *differentially-encoded quadrature phase-shift-keying*. The frequency of the carrier is nine times the bit-rate of 728 kbit s^{-1}.

A pair of bits may have one of four states (00, 01, 10, 11), and each of these four states corresponds to one of four possible changes to the phase of the carrier (see Figure 6.3). It must be stressed that this differs from QPSK, which encodes pairs of bits

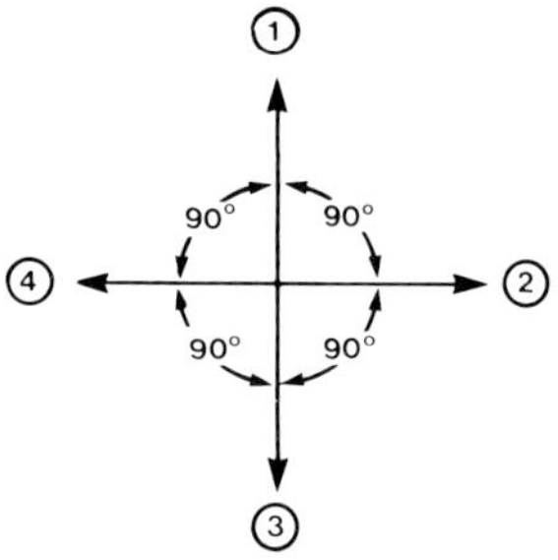

Figure 5(a): Rest-states of carrier phase.

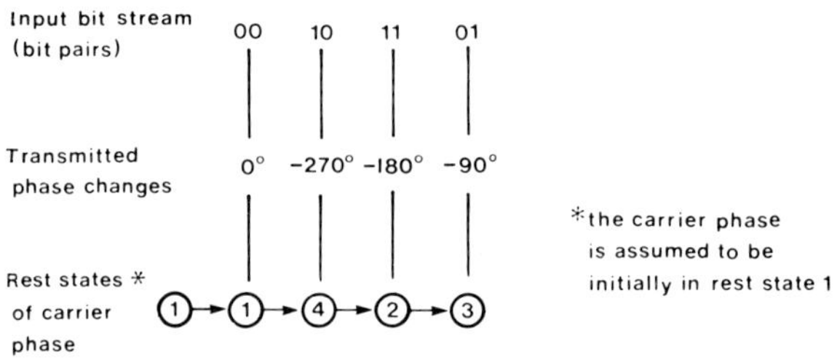

Figure 6.3 DQPSK carrier phase changes related to NICAM 728 bit pairs (Courtesy BBC)

by the *absolute* phase of the carrier, rather than by the *change* in its phase relative to the last state. DQPSK has the advantage that no phase reference is required.

The carrier, thus modulated, is set at a level of − 20 dB in relation to the amplitude of the vision carrier, and is thus 10 dB below the amplitude of the FM sound carrier (see Figure 6.4).

In PAL systems B and G, the frequency of the digital sound carrier is adjusted to be somewhat lower than 6.552 MHz, in order not to interfere with the upper adjacent vision carrier. This is made feasible because the frequency of the mono FM carrier is set at 5.5 MHz in these systems.

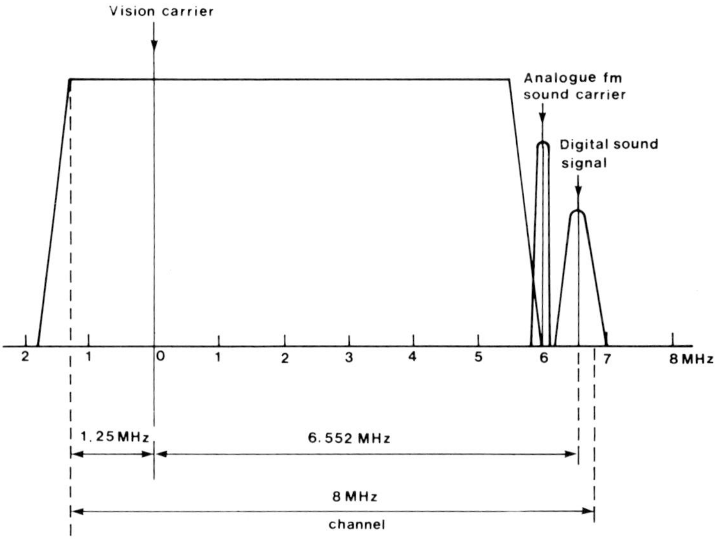

Figure 6.4 Transmitted spectrum including NICAM 728 carrier (Courtesy BBC)

Performance of NICAM 728

The use of a digital method to transmit stereo sound offers high quality, low crosstalk and freedom from interference. It also leaves the existing mono carrier intact. Tests of experimental NICAM systems have shown that the sound channel remains intact in conditions under which the quality of the picture signal is judged to be very poor indeed. In cases where picture signal-to-noise ratios have been below 20 dB, the NICAM sound channel has not failed, and in such a case the picture would be judged virtually un-watchable.

NICAM transmissions appear to stand up well to conditions where multi-path reflections are prevalent, and have the great advantage that the sound quality remains virtually unchanged as interference increases. Close to the limits of noise tolerance, where the error correction system may resort to error-concealment or interpolation (instead of perfect correction), quality may be marginally reduced, but the system will normally mute before sound quality falls below the level at which it becomes unacceptable.

NICAM sound in satellite systems

NICAM 728 has been designed for maximum commonality with the method of sound transmission used in the MAC (multiplexed analogue components) system intended for satellite broadcasting. D-MAC allows for companded digital audio as well as for linearly-coded audio, and it does not specify the use of NICAM alone as the universal coding method. Nonetheless, it seems likely that NICAM may become one of the coding methods adopted in parts of the world, although the overall frame structure is not expected to be quite the same as NICAM 728, having a data rate of 751 kbit s^{-1} instead of 728 kbit s^{-1}. Thus it is intended that the same chip-set which has been designed to decode NICAM 728 domestically would be able to handle the sound from some MAC satellite broadcasts as well.

BTSC multi-channel sound transmission

The system adopted for multi-channel sound transmission with terrestrial television in the United States (also to be adopted in Canada) is an analogue system employing noise reduction. It is based on similar principles to the Zenith stereo multiplex method used in high quality FM radio transmission, and the noise reduction employed is a variation on the well-known dbx system, optimized for television use. In the NTSC transmission system there is a certain amount of unused channel space around the existing mono FM carrier, which allows for extra deviation of this carrier due to the addition of sub-carriers to the baseband signal.

Encoding of stereo information

In the BTSC system (sometimes referred to as MTS, for Multi-channel Television Sound), the monophonic sound channel is retained for compatibility with existing receivers. This channel therefore retains its characteristic as a frequency-modulated carrier with ± 25kHz deviation and a frequency response of 50 Hz–15 kHz, and forms the A + B sum (or M signal) of the stereo broadcast. The mono signal is subject to 75 μs pre-emphasis to reduce the effect of noise.

As in the Zenith GE stereo multiplex, a sub-carrier is added to the baseband before frequency modulation, which carries the stereo difference information (the (A–B) or S signal). This sub-carrier is centred at 31.468 kHz, which is twice the horizontal line frequency of the NTSC television system (15.734 kHz), and the modulation method used is double-sideband suppressed-carrier (DSBSC). The frequency response of the difference channel is also 50 Hz to 15 kHz. Because of the increased noise which usually results in the difference channel with this method of stereo encoding, noise reduction is applied to the audio signal (see below).

As indicated in Chapter 2, inter-channel gain and phase errors have a significant effect on stereo signals, the effect on MS signals being a worsening of crosstalk with increased error. If the stereo separation is not to be compromised it is vital that the compansion system tracks very closely between the receiver and transmitter, and that the different modulation methods used for the M and S signals do not result in phase errors.

BTSC noise reduction

Based on the principles of the dbx noise reduction system used in recording systems, the BTSC system incorporates both fixed pre-emphasis with a number of time constants, wideband compression, and frequency-dependent compression. As with most compansion systems, it relies on the ear's tendency to ignore noise in the presence of a signal within the same band. The block diagram in Figure 6.5 shows the main processing areas involved.

The LF bandpass filter operates up to just over 2 kHz, providing low-to-mid frequency compression, and the HF bandpass filter rises from 500 Hz, peaking at 10 kHz, to provide high

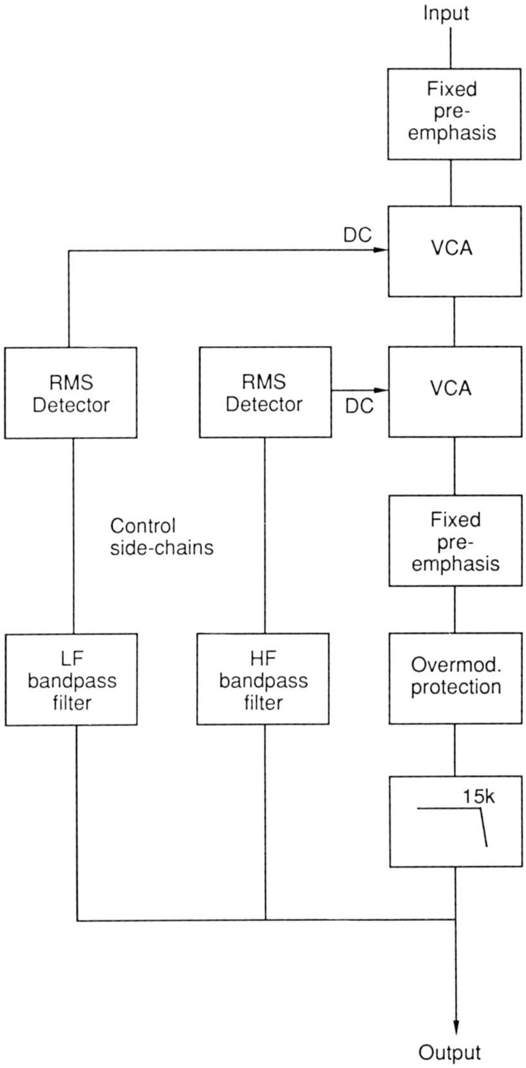

Figure 6.5 Block diagram of BTSC noise reduction encoder

frequency compression. The system also includes protection against over-modulation of the difference channel.

With this type of noise reduction used on the difference channel, together with the additional main carrier deviation, it is claimed that the signal-to-noise performance is no worse in stereo than it is in mono.

Secondary audio program (SAP)

A further sub-carrier, centred at five times the TV horizontal line frequency (78.67 kHz) is used for a secondary audio program, which may be used to carry a second language or other additional material. Its bandwidth is slightly lower than the main channel, being only 50 Hz–10 kHz, and, like the stereo difference channel, it is subject to noise reduction. Unlike the difference channel, the SAP is frequency modulated onto its sub-carrier, with a peak deviation of ± 10 kHz.

Non-public channel

A final FM sub-carrier is centred on six-and-a-half times line frequency (102.3 kHz), and this is intended for 'non-public' purposes, such as the carrying of data, or as a low-quality communications channel. For speech purposes the bandwidth is 50 Hz–3.4 kHz, using 150 μs pre-emphasis, and for data purposes it will support rates from 0 to 1500 Hz with no pre-emphasis. The deviation of the sub-carrier is ± 3.5 kHz maximum.

Resulting baseband spectrum

Figure 6.6 shows the baseband spectrum that results from the addition of the sub-carriers, and it can be seen that the amplitudes of the sub-carriers are set such that each contributes a certain amount to the total deviation of the main carrier, which is allowed to have a peak deviation of up to 73 kHz (as opposed to the 25 kHz in mono operation). This is somewhat different to the situation in Zenith GE stereo in which the total deviation of the main carrier remains unchanged at 75 kHz (Band II FM, UK), even with the addition of the stereo difference sub-carrier, leading to a poorer S/N ratio. Although the modulation index of the BTSC main carrier in stereo will be slightly lower than it was mono (leading to a poorer overall S/N ratio in the demodulated signals), this is compensated for by the use of noise reduction in higher-frequency sub-carrier channels, as indicated above.

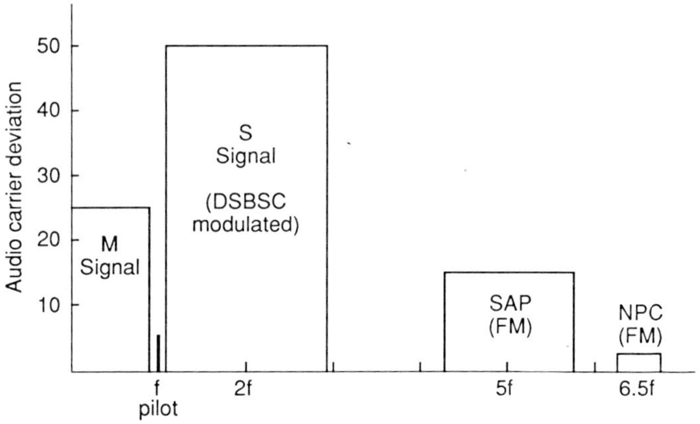

Figure 6.6 BTSC audio baseband spectrum. f = NTSC horizontal line frequency (15.734 kHz)

The total contribution which each sub-carrier makes to the deviation of the main carrier is given below:

Main channel (A + B)	25 kHz
Pilot (15.734 kHz sine wave)	5 kHz
Difference channel (A–B with NR)	50 kHz
SAP (Mono with NR)	15 kHz
Non-public channel	3 kHz

It is stipulated that the total *instantaneous* carrier deviation due to the main channel and the stereo difference channel should not exceed 50 kHz, and in practice this is achieved for most program material due to the compression of the difference channel and the nature of the stereo difference information.

A pilot signal is added at the NTSC horizontal line frequency (15.734 kHz), to act as a reference for decoding, and this appears below the lower sideband of the difference channel, and just above the main audio channel. There is not much space for it here, and thus steep filters are required both to roll off the lower DSBSC sideband and the audio baseband. Additionally, it is important that the lower DSBSC sideband does not alias with the audio baseband.

In order that response errors do not arise between the sum and difference channels (resulting in crosstalk) their roll-off filters must be matched very closely. In practice, channel separation is around 30–35 dB at low-mid frequencies, falling to 20–25 dB at HF. This is adequate for subjective separation of left and right channels (see Chapter 1), but is not as good as that obtained from a digital system. In the BTSC system, the left and right channels of the main program would be unlikely to be used for separate languages or program material, as there is a separate sub-carrier for such purposes, thus the lack of separation is not a major problem.

Dolby ADM system

A contender for the UK stereo sound transmission standard, and adopted in Australia for stereo sound on Direct Broadcast Satellites (DBS), Dolby Laboratories' Adaptive Delta-Modulation system is a means of digitally encoding audio information using a high sampling rate but small number of bits per sample. Such a method has certain advantages: principally that decoders may be relatively cheap, and that the traditional low-pass filter required in digital convertors need not be nearly as steep or precise as in conventional PCM systems due to the high sampling rate. Additionally, there is considerable flexibility in the transmitted data rate to suit a number of applications, having a minimal effect on perceived audio quality.

Principles of operation

The process of delta modulation is already well documented, being a means of using essentially a one-bit code to indicate changes up or down in the level of the audio signal. Provided that such changes are signalled regularly enough, the perceived quality of such a system may be as good as that of linearly-encoded PCM. The sampling rate required to accommodate high frequency, high level audio signals at similar quality to linear PCM is considerably greater than that of a system like NICAM (see above), based on the laws of information transfer which can be summed up by the generalization that the same amount of information can be transferred using 16-bit samples at a sampling rate (n), as can be transferred using 1-bit samples at a sampling rate of ($16n$). The high sampling rate has the benefit of allowing a shallow low-pass filter to roll-off components above the audio baseband, as aliasing is much less likely to occur.

Adaptive Delta Modulation is used for broadcast purposes because it is difficult to get a wide enough dynamic range out of a simple delta modulation system whilst keeping the bit-rate within reasonable limits. Effectively it is a non-linear system, in that the step-size (change in audio level) represented by each bit can be altered using an additional control signal, this being adapted to suit the programme material such that programme-modulated noise is kept to a minimum. With ADM, the choice of step-size is always a compromise between it being too small (leading to distortion related to slew-rate limiting), and it being too large (leading to programme-modulated noise), thus the key to the adaptive principle is in the effective prediction of the correct step size.

As the sampling rate is high, quantization noise is spread over a wide frequency band, much of it outside the audible range, and thus it is possible to employ noise-shaping techniques to remove in-band noise at the expense of noise at higher frequencies (out-of-band). Complementary to this, variable pre-emphasis is used, based on the spectral content of the audio signal, to optimize the signal-to-quantizing-noise ratio for a wide variety of different input signals and step-sizes. A subsidiary data stream is generated to control the de-emphasis at the decoder.

The block diagram in Figure 6.7 indicates the main processing hardware of such an ADM system, and it may be seen that it is essentially a feed-forward system whereby the input audio signal is delayed to allow for forward-looking prediction of the correct step-size and emphasis. The total delay in this particular system, amounting to some 20 ms, might become significant if synchronization with a picture was concerned.

Typically, it is possible to achieve good audio quality in a broadcast application with an audio data rate of between 200 and 350 kbit s^{-1} per channel, whilst the control signals for step-size and emphasis are at a lower rate of around 8 kbit s^{-1} each. This matches closely with the data rate used in the NICAM 728 system, and thus would lend itself to adoption in similar situations where limited band-space is a problem.

Transmitting ADM

Since the data rate is similar to NICAM 728, it is reasonable to assume that a similar modulation method might be used in order to transmit ADM sound alongside terrestrial TV signals, and this

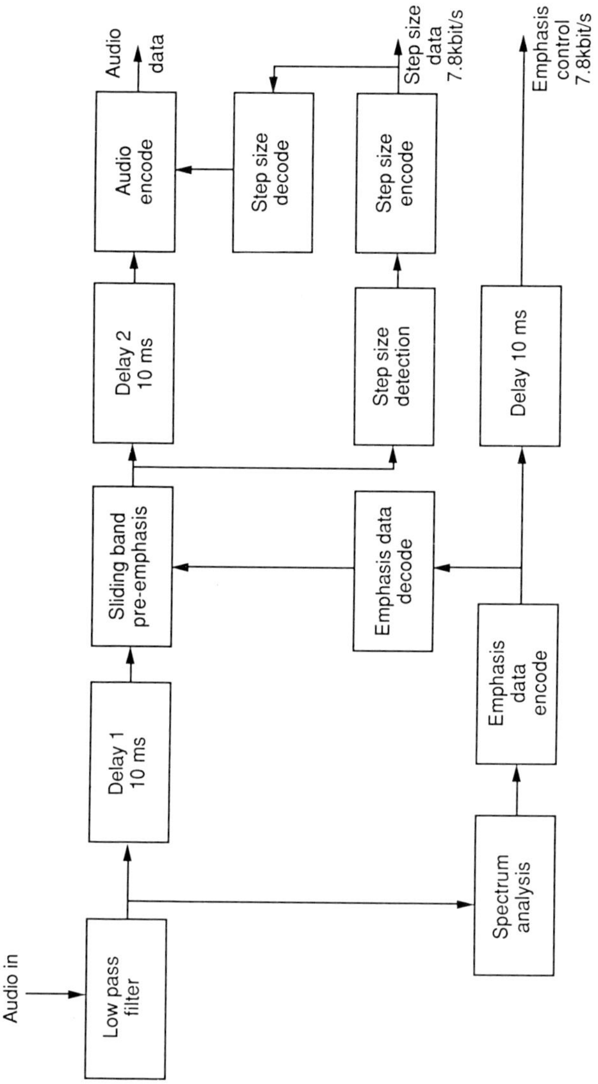

Figure 6.7 Block diagram of Dolby ADM system encoder (Courtesy Dolby)

has been suggested by Dolby Labs. In the event, NICAM has been adopted by the EBU as its de facto digital standard for stereo television sound, although the ADM system has found favour in Australia for use with the B-MAC DBS (Direct Broadcast Satellite) system.

In the Australian DBS system, six channels of ADM-encoded sound are transmitted at an audio data rate of 204 kbit s^{-1} per channel, with 13-bit packets sent once every TV line. The additional sound capacity is used for the transmission of two digital stereo radio channels. Furthermore, Australian independent TV stations have recently adopted ADM for use in a professional distribution system handling a picture and up to four high quality audio channels, with audio on a single sub-carrier above the vision signal.

Further stereo transmission systems

Dual-carrier FM

A means of transmitting stereo information by the addition of a second FM carrier, slightly higher in frequency than the mono carrier, has been adopted in West Germany. The system was developed jointly with Sweden, although it is now more likely that the Scandinavian countries will move towards the use of a NICAM-based system.

Somewhat unusually, although for good reasons, sum and difference signals are not employed in the conventional manner for transmission. In the dual-carrier system, the mono compatible carrier is indeed (A + B) or M, but the second carrier, spaced 24 kHz above this, holds only the right channel, or B. It may be appreciated that separate A and B signals can still be derived at the receiver by subtracting B from the mono sum, giving A. In bi-lingual operation, each carrier takes a separate language, and the separation is good. As discussed earlier, any sum and difference method is likely to suffer increased crosstalk in the face of level and phase errors, but the use of separate carriers for each language avoids this problem.

In order to allow receiver switching between bi-lingual and stereo modes, a control signal is modulated onto a 55 kHz AM sub-carrier which is included in the baseband of the second FM carrier.

This system has proved successful in the period of operation in Germany since 1981, although it does not offer the improved sound quality available with a digital system, nor does it allow for any additional information to be broadcast along with the television signal. Other countries which have used or considered the adoption of dual-FM stereo sound are Australia (for terrestrial broadcasts), Austria, Italy, Holland, and Switzerland, although the European countries are being encouraged by the EBU to adopt NICAM.

FM/FM

Japan which has had stereo sound longer than any other country, has adopted a system which dates back to an experimental Motorola design from the 1960s, although this has changed significantly since its early days. Today's system was introduced in 1978 and involves the use of an FM carrier for mono sound, incorporating a stereo FM sub-carrier within the un-modulated baseband (similarly to the DSBSC difference sub-carrier in Zenith stereo; see above). The sub-carrier, at 31.47 kHz, carries the (A–B) or S signal in stereo mode, or a second language in bilingual mode, the mode being switched by a control signal modulated onto another sub-carrier.

Since the additional sub-carrier and its sidebands must be accommodated within the baseband (which also contains audio), the deviation of this sub-carrier cannot be great otherwise too large a number of sidebands would be produced, interfering with the mono signal. Limiting the deviation of the sub-carrier leads to higher noise in the difference or second language channel, as well as possible distortion. In order that the modulation index of the FM sub-carrier does not become unreasonably low (this being the main factor dictating the noise performance) the frequency response of the audio signal on this channel is also limited.

Further reading

BENSBERG, G. J. and EMMETT, J. R. (1988) Dual channel sound in the ITV network. In *Proceedings of the Institute of Acoustics*, **10.7**, 75–79 (ed. R. Lawrence), Institute of Acoustics

BBC (1988) NICAM 728: specification for two additional digital sound channels with System I television. *BBC Engineering Information*, August

BKSTS (1989) Stereo Sound for Picture—report on a BKSTS seminar of 6 November 1988, *Image Technology*, February, pp 52–57

BORWICK, J. (ed.) (1988) *Sound Recording Practice*, Oxford University Press, Oxford

ELY, S. R. (1983) Experimental digital stereo sound with terrestrial television: field tests from Wenvoe, October 1983. *BBC Engineering Information*, December

ELY, S. R. (1986) Progress and international aspects of digital stereo sound for terrestrial television. In *Proceedings of the International Broadcasting Convention, 1986*, 138–141, Institute of Electrical Engineers, London

FISHER, J. (1986) Stereo and multi-lingual television sound broadcasting. *International Broadcast Engineer*, July, 48–55

FORSHAY, S. E. (1987) Economical digital audio for broadcasting. Reprinted from *Electronics and Wireless World*, March, Reed Business Publishing, London

HARDING, W. B. (1984) Audio coding methods to extend the DBS sound coverage. In *Proceedings of the International Broadcasting Convention, 1984*, 185–187, Institute of Electrical Engineers, London

HOLDER, J. E., SPENCELEY, N. M., and CLEMENTSON, C. S. (1984) A two-channel sound-in-syncs transmission system. In *Proceedings of the International Broadcasting Convention, 1984*, 345–348, Institute of Electrical Engineers, London

HOFFNER, R. (1987) Multi-channel television sound broadcasting in the United States. *Journal Audio Engineering Society*, **35.9**, September, 660–665

JOHNSON, M. F. (1988) Another year on with stereo sound at BBC TV. In *Proceedings of the Institute of Acoustics*, **10.7**, 61–64 (ed. R. Lawrence), Institute of Acoustics

NISBETT, A. (1989) *The Use of Microphones*, 3rd edn, Focal Press, London and Boston

SESTERIKOV, B. V. (1986) The transmission of multi-channel television sound. In *Proceedings of the International Broadcasting Convention, 1986*, pp 136–137, Institute of Electrical Engineers, London

TAYLOR, N. K. (1984) Digital Audio in Broadcasting—the opportunities and challenges. In *Proceedings of the International Broadcasting Convention, 1984*, pp 358–360, Institute of Electrical Engineers, London

WATKINSON, J. (1988) *The Art of Digital Audio*, Focal Press, London and Boston

WHITE, J. (1988) Stereophonic sound—will it work for television? In *Proceedings of the Institute of Acoustics*, **10.7**, pp 67–74 (ed. R. Lawrence), Institute of Acoustics

Glossary of Terms

A-signal the term conventionally used for the left hand channel of a stereo pair, or channel one of a dual mono pair. It takes the colour red in wiring and metering conventions.
AB pair a stereo pair of signals existing in the format which can be fed directly to loudspeakers (left–right), or a pair of microphones orientated so as to pick up stereo sound in this format.
AES/EBU a digital interface standard for the transfer of two channels of audio information over a single serial link.
ADM Adaptive Delta Modulation is a digital system developed by Dolby Laboratories for the transmission of high quality digital audio at a data rate between approximately 200 and 320 kbit s^{-1} per channel.
Ambisonics a system which may be used to record and reproduce surround-sound information, adopted in the Calrec Soundfield microphone
B-signal the term conventionally used for the right hand channel of a stereo pair, or channel two of a dual mono pair. It takes the colour green in wiring and metering conventions.
BBC the British Broadcasting Corporation.
BTSC Broadcast Television Standards Committee: the analogue transmission system used in the USA and Canada for stereo sound.
Baseband the frequency spectrum of a signal before modulation.
Betacam SP a cassette-based video recording format which offers two high quality FM sound tracks in addition to two analogue tracks with Dolby C noise reduction.

Binaural a recording technique which places microphones in the positions of human ears, sometimes using a dummy head, designed to be reproduced on headphones in an attempt to simulate the original soundfield.

Blumlein Alan Blumlein, the man who is considered to be the 'father' of stereo, sometimes used in reference to a particular microphone technique involving a crossed-pair of co-incident microphones which he pioneered.

C-format a reel-to-reel video format offering two linear audio channels plus a timecode track. Recent implementation of two digital audio tracks is useful for stereo.

Centre timecode the incorporation of a timecode track in the guardband between the tracks of a NAB quarter inch two-track audio recorder.

Codec a coder/decoder pair in a digital system.

Comb-filtering the effect which arises when two similar signals are added together which have a slight phase difference, resulting in notches in the frequency spectrum of the summed signal due to cancellation.

Companding compressing and expanding the dynamic range of an audio signal so as to reduce the effects of noise in a transmission channel.

Crosstalk the breakthrough of one audio channel to another, resulting in a loss of separation.

D-1 and D-2 these are digital video recording formats which also carry four channels of high quality digital audio.

DBS Direct Broadcast Satellite.

Deviation the instantaneous change in frequency of an FM carrier which is related to the amplitude of the modulating signal. The maximum deviation allowed varies with the system.

De-emphasis the implementation of a filter characteristic which complements that used in **pre-emphasis**, so as to restore the spectral content of an audio signal to its original form whilst reducing unwanted noise.

Decca Tree an arrangement of three omni-directional microphones with specific distances between them, to record in stereo.

Decoding the re-instatement of a digital audio signal to its original form, correcting errors and un-scrambling the data (used after encoding).

Direct field the acoustic field, near to a sound source, where the direct information from the source predominates over reverberant information.

DQPSK Differential Quadrature Phase Shift Keying: a method of modulating digital data onto a carrier by changing the *relative* phase of the carrier in steps of ninety degrees, corresponding to the state of pairs of data bits (used in NICAM 728).

DSIS dual sound-in-syncs (see **sound-in-syncs**).

Dual-FM the method by which two FM sound carriers, separated by 24 kHz, are used to transmit stereo information, one carrying A + B, the other carrying B only (used in West Germany and others).

EBU the European Broadcasting Union.

Emphasis see **Pre-emphasis**.

Encoding the formation of a sequence of digital audio data into a pattern suitable for transmission or recording, perhaps involving the addition of error correction and additional data.

EQ equalization; a means of changing the frequency response of an audio signal.

FM frequency modulation: a means of modulating a signal onto a carrier by using the amplitude of the signal to vary the frequency of the carrier.

FM/FM the method of encoding stereo information onto a single FM sound carrier by incorporating an FM sub-carrier in the baseband to handle the difference signal or second language (used in Japan).

Frame a definable block of information which is repeated, such as a television frame (a single still picture), or a digital audio frame (a block of data containing a given number of samples).

Gain-ranging a method of indicating the original level of a compressed block of samples to a digital decoder, such that they can be restored to the correct gain range (used in NICAM 728 transmission systems).

Ganging the system of linking controls together such that one knob or control signal acts on more than one signal chain.

Haas Effect the psycho-acoustic effect of perceiving directionality by locking onto the first arrival of a sound source (within certain limits).

IBA Independent Broadcasting Authority; the controlling organization for Independent Television in the UK.

MAC Multiplexed Analogue Components; a transmission standard for satellite broadcasting.

Matrix an electrical network designed to convert signals from one format to another; from AB to MS format, for example.

MS pair a stereo pair of signals existing in the form of a mono sum and a difference channel (A + B) and (A-B), or a pair of microphones orientated so as to pick up stereo signals in this format.

Mono compatibility a term which refers to the degree of compatibility between a stereo pair of signals and its mono sum. The mono compatibility may be affected by such factors as phase errors and level errors between the A and B signals. Such compatibility is regarded as important since many listeners will still receive in mono.

M signal the monophonic sum of A and B signals in a stereo pair, sometimes corrected in level using between 3 and 6 dB of attenuation to compensate for the rise in level which occurs in summation (may be referred to as Main or Middle).

Monitor may refer either to the audio monitor (the loudspeaker) or the TV monitor (the screen), or the main listening output of a mixing desk.

M-II format a cassette-based video recording format which offers two channels of digital sound in addition to analogue and FM sound tracks.

Near field monitors monitor loudspeakers which reside close to the balance engineer, such that he is within a region where direct sound predominates. These may be useful in rooms whose acoustics are poor.

NICAM Near Instantaneously Companded Audio Multiplex: a digital audio system developed by the BBC which involves the reduction of audio resolution from fourteen bits to ten bits for transmission at a low data rate of around 320 kbit s^{-1} per channel, with subsequent reconstruction on decoding. The sampling rate is 32 kHz.

NICAM 728 a digital stereo transmission system used in the UK, and recommended by the EBU for use with PAL TV systems, operating at a data rate of 728 kbit s^{-1}, and carrying two channels of audio plus data.

NICAM 3 a digital audio communications system developed by the BBC, using the principles of NICAM coding adopted also in NICAM 728.

Normal incidence the term used to describe signals arriving on-axis to the front of a microphone or stereo pair.

NTSC the National Television Standards Committee; used in reference to the standard used for 525-line colour transmission in the US, Japan and others.

Out-of-phase the term used to describe the situation when two audio signals are 180° out-of-phase with each other.
PAL Phase Alternate Lines; used in reference to the 625-line colour transmission standard used in Europe and elsewhere. PAL-I, B and G refer to variations in the PAL standard in different countries.
Pan control a control found on a mixer which places a mono sound signal somewhere between left and right in a stereo image by varying the relative levels between the channels.
Phase meter a meter which indicates the degree of phase difference between two signals.
PPM peak programme meter: a mechanical meter with specified ballistics which responds to peak levels, having a scale from 1 to 7. Stereo PPMs have two needles coloured red and green. M and S PPMs have white and yellow needles.
Pre-emphasis a means of boosting the gain of an audio signal over a particular frequency band before recording or transmission such that when the signal is **de-emphasized** at the receiver the signal is returned to its original level whilst any noise from the transmission channel is reduced.
Programme modulated noise noise which arises only in the presence of an audio signal, and may be due to errors in the quantization of a digital audio signal.
Quantization the process of measuring the amplitude of an audio signal sample and assigning to it a binary number which is as close as possible to its true level in the range of numbers available in a digital system.
QPSK Quadrature Phase Shift Keying: a means of modulating digital data onto a carrier by changing its phase in steps of 90°.
Quaternary coding the coding of pairs of binary digits such that they correspond to four discrete voltage levels (used in sound-in-syncs).
S signal a signal derived by subtracting the B signal of a stereo pair from the A signal (A-B), sometimes corrected in level by attenuation of between 3 and 6 dB to match the attenuation of the M signal. Otherwise known as the 'difference' or 'side' signal.
Sampling rate this is the number of times per second at which an audio signal is sampled in a digital system. In broadcast transmission systems this is 32 kHz, and in professional audio systems it is 48 kHz or 44.1 kHz.
Sample-rate conversion a method by which digital audio

signals are converted from one rate to another without passing through the analogue domain.

Scaling factor the digital code word sent with a block of compressed samples in a NICAM system to indicate the correct level of the sample block.

Stereo pair a pair of signals which are the left and right channels of a stereo programme (this might also refer to the pair of M and S channels). Sometimes used to refer to a stereo pair of microphones.

Sync pulse a portion of the television signal waveform which synchronises the start of each line of the picture, sometimes used for the insertion of digital audio data (see **sound-in-syncs**).

Sound-in-Syncs a method of inserting digital audio data into the line sync pulses of a television waveform so that audio information can be carried over the same link as a picture. Recently modified to carry stereo information.

Time Delay Spectrometry an acoustic analysis tool which enables the user to see how the sound energy in a room is distributed in time, frequency and space.

Vector scope a metering device which displays a trace indicating the relative phases and levels of A and B signals. The shape of the vector display can be interpreted to indicate any problems with the stereo signal.

Variable pre-emphasis pre-emphasis whose characteristic is changed according to the spectral content of the programme signal.

Width control a control which may be found on some stereo equipment such as mixing consoles, used to change the apparent width of the sound-stage. It is in fact an 'S-gain' control acting on an MS pair.

Index

References to the glossary are indicated by 'g', e.g. AB pair, 96g

Stereo Sound for Television

Stereo Sound for Television covers every aspect of stereo sound, from perception and psycho-acoustics, through microphone techniques and stereo broadcast equipment, to distribution and transmission systems.

All currently adopted transmission systems are discussed; NICAM 728, BTSC, dual FM system, FM/FM system and the Dolby ADM system, together with coverage of the AES/EBU digital interface and stereo sound-in-syncs distribution.

With a foreword by Jeff Baker, Head of BBC Television Sound, *Stereo Sound for Television* will be an invaluable aid for operators, engineers and management, as well as students intending to enter the field of TV broadcasting.

Francis Rumsey is a lecturer in Acoustics and Sound Recording at the University of Surrey. He has contributed articles to many of the leading audio and broadcast magazines and has designed and lectured on stereo sound training courses for the Independent Television Association.

ISBN 0-240-51288-X

9 780240 512884

Focal Press
Borough Green, Sevenoaks
Kent, TN15 8PH, England

80 Montvale Avenue, Stoneham
MA 02180, USA